PROFESSIONAL EXPRESSION

PROFESSIONAL EXPRESSION

TO ORGANIZE, WRITE, AND MANAGE TECHNICAL COMMUNICATION

M. D. MORRIS, PE

F. ASCE; F. STC

MOMENTUM PRESS, LLC, NEW JERSEY

Professional Expression
Copyright © Momentum Press®, LLC, 2009

All rights reserved. No part of this publication may be reproduced, stored in a retrieval system, or transmitted in any form or by any means—electronic, mechanical, photocopy, recording or any other except for brief quotations, not to exceed 400 words, without the prior permission of the publisher.

First published in 2009 by
Momentum Press®, LLC
222 East 46th Street, New York, NY 10017
www.momentumpress.net

ISBN-13: 978-1-60650-071-2 (paperback)
ISBN-10: 1-60650-071-6 (paperback)

ISBN-13: 978-1-60650-072-9 (e-book)
ISBN-10: 1-60650-072-4 (e-book)

DOI forthcoming

Cover Design by Jonathan Pennell
Interior Design by Scribe, Inc.

First Edition June 2009

10 9 8 7 6 5 4 3 2 1

Printed in the United States of America.

for
Abraham Lincoln Morris
(1878–1958)
Mentor, Friend, Father
He taught me to design word pictures.

CONTENTS

Foreword by Francis J. Lombardi, PE	ix
Disclaimer	xi
Preface	xiii
Acknowledgments	xv
1. Basics	1
2. Organization and Development	13
3. Writing Technique	39
4. Characteristics, Refinements, and Style	43
5. Descriptions	53
6. Comparisons	63
7. Motion	73
8. Short Forms	81
9. Formats	91
10. Expanding Outline and Text	103
11. Focus Analysis	107
12. Alternative Specialized Formats	111
13. Oral Presentations	115
14. Authorship	123
15. Articles	127
16. Books	137
17. Critique, Discussion, Rebuttal	149
18. Letters	155
19. Appendices	163
20. References and Bibliography	191
Postlude	195
Index	197

FOREWORD

SEEK FIRST TO UNDERSTAND, AND THEN TO BE UNDERSTOOD.

—Stephen Covey, *The Seven Habits of Highly Effective People*

How many times have we heard that the explanation or excuse for missing a goal or milestone was "poor communication"? Communication does not have to cause problems; communication can prevent them. *Professional Expression* by M. D. Morris, PE, shows you how to turn communication into an asset rather than a liability. Remember: The single greatest factor for advancement is your ability to convey your ideas.

Good communication involves taking what you've heard and learned through active listening and focused research, and then presenting that complex information in a clear and concise manner that can be understood easily by any audience. This book will help you do just that. It contains many proven approaches to developing and improving your communications skills. You'll learn helpful methods for organization and development, efficient writing techniques for report and proposal formats, and how to deliver successful oral presentations. Those chapters are intended to help you understand why each of those components plays a part in successful communication, and how they work together to heighten your communication competence.

Let's face it, most technical people get their education using numbers, not words. This book will help you choose your words carefully and use them in a practical way to improve your professional career.

Let me recommend that you read the book in its entirety first. Reread it to hone in on problem areas. Then practice, practice, and practice some more!

M. D. Morris, PE, has been teaching this course to our technical staff for the past six years, and it has made a significant difference in helping their level of technical communication become more professional and successful. As someone who has benefited firsthand from this new way to approach any communication task, believe me: It's a proven guide!

Francis J. Lombardi, PE
Chief Engineer
The Port Authority of New York and New Jersey

DISCLAIMER

I write this volume in full agreement with the fundamental fact of equality of women in the professional workforce. But consistently to write "he or she," "he/she," "his or hers," "his/hers" is inefficient, uneconomical, and distracting to the smooth flow of this narrative. To save time and money, I use the terms "he" and "his" as acronyms or symbols to be understood by all to mean "men, women, everyone," just as you would read the symbol π (pi) to mean 3.14159....

All this material is original and may not be reproduced in any form without written permission. Attributed short quotes are permissible. I use real examples of good and bad writing to illustrate the point at hand. For the most part those have been produced in my classes since the beginning and are used with the full knowledge and consent of their writers. The positive ones are fully attributed. The bad ones remain anonymous to spare their originators any possible embarrassment.

Over the years since I first developed them in 1965, several parts of this text have appeared in numerous other publications as articles, editorials, contributed book chapters, or quotations. All stemmed originally from my own thoughts and have been used in all my teaching.

PREFACE

This book is a practical expansion of contemporary plain professional English as it is spoken and written in the technical and administrative worlds. It is *not* an academic text after college English 101, nor does it deal with the language of current popular culture. This is a guide for all people whose livelihoods depend on their understanding, and their being understood.

Actually, I am passing along to you the strongest valid points from the experience I've gleaned after half a century of published writing, and editing, and more than forty years of teaching adult working practitioners.

Early in life my sage father taught me, "Profit from someone else's experience. If you can't, then profit from your own." Don't go through life wearing pairs of horse blinders and earplugs, caught up in the herd.

You need this system for your professional expression, to stay ahead of those who don't yet have it. Here is a new way to approach and accomplish *every* communication task. You no longer need to stare at a blank sheet of paper or a blank computer screen. Here, for all your writing and speaking assignments, you will find out where to start, and how to go.

Instead of the archaic writing procedure that has been taught since technology became a field apart from arts and letters, try a new protocol bereft of those outmoded limits and flaws. Replace the old routine of concocting an outline (of course, within the limits of your own knowledge), plugging in researched material, and writing. Then the writing would consist of an introduction, a body (facts and opinions), and a conclusion. All written in the impersonal third person, in the passive voice. All wrong!

Where in all that is the place for things *outside* your knowledge purview? You don't know what you don't know. This new approach can reveal the blind spots in your vista. Your written product via this method will read easier and more understandably, and thus be more useful to your target audience. In this world there are absolutely no absolutes.

From many diverse good ideas, I generated this procedure by amalgamating and organizing them for the first semester of my teaching career. That was in the evening school of continuing adult education at The Cooper Union in New York. My first son Gregory was born the night of my first (twenty-six-student) class. He has now become an independent writer, having gone past being editor-in-chief of a major industrial trade magazine.

Because this method worked the first time, I used it again the next semester, then in more than 800 short- and full-term credit courses with some 17,000 students. For professionals on the job, I condensed the high points into a short, in-house course of three eight-hour days that has earned excellent evaluations. Among the short packages were stints at NASA, Los Alamos, and many other federal installations, as well as for major industry and professional organizations. Full-term credit courses met with rewarding results at The Cooper Union, Columbia University Civil Engineering, and the MBA program at Cornell University.

There are no busywork exercises in here. This is a pragmatic tome written for use as a guide in the working world. Follow these exercises in parallel with your own job at hand. You may not get the results you seek on the first try, but with a little practice you shall. I've taught this in English in Japan, and in Spanish in the Caribbean; it worked well in both places.

Please note that although the firm rule of the road here is to keep it simple, as my old lawyer friend (when pushed about the opacity of his "legalese") would respond, "We never endeavor, we only try!" Herein, instead of the correct commonplace words, I deliberately use many multisyllabic words to get you to use your dictionary and broaden your horizons. Also, I suggest you read this all through once to "case the place," then keep it close at hand for ready reference.

Get into this natural, common-sense communication system with an open mind: it works with or without your computer. You'll find it a positive route to making your professional expression easier, more successful, and more enjoyable.

M. D. Morris, PE
Ithaca, NY
2008

ESSAYONS[1]

NOTE

1. From the Latin, "We shall try." The motto of the Corps of Engineers, U.S. Army.

ACKNOWLEDGMENTS

Although I am the originator of this work (concept through publication), I didn't arrive at this datum without aid along the way. Were I to thank all who helped, I'd need a full chapter, but eight persons among them generated significant effects on my writing life! I am grateful to every one of them. I name these eight because singly each appeared at a critical point to keep me on track and going ahead:

Frances Aigeltinger of P.S.166, who undertook to nurture this nascent narrator.

Harry Shefter of Stuyvesant High School, who taught me useful English for four of those eight terms.

Charles Kenneth Thomas, professor of public speaking at Cornell University (later department chairman), who, despite my resistance, took my New York "dead end" brogue and forged it into articulate discourse.

Dr. Ralph Palmer Agnew, Cornell professor of mathematics, who inspired me to realize I could be a teacher, and a writer.

William H. Quirk, PE, editor of *Contractors and Engineers Monthly*, who early on helpfully rescued my writing career at its lowest starting point.

Irwin Forman, PE, editor of *Ingeniería Internacional Construcción*, who confidently bought everything I filed from Latin America.

Dr. Johnson E. Fairchild of The Cooper Union, who invited me to teach my "good material" in his adult education evening school curriculum.

Anita Diamant Berke, a literary agent at the Overseas Press Club, who secured a publisher's contract for me to write *Okinawa*,[1] my first book.

Probably none of them is alive today, but unlike the "Wamba-dan" the turtle of old Chinese philosophy, I am not "one who has forgotten the eight."

I cannot overlook two more people, in on this finished effort: Joel Stein, publisher for Momentum Press, LLC, who had the astute good judgment to contract this project; and Elaine Meredith Guidero, who typed and retyped every manuscript page, in addition to spot research, and copy editing. Thanks to you all.

DO IT, DON'T DON'T IT![2]

NOTES

1. New York: Hawthorn Books, 1970.
2. Morton W. Jacobs, Esq. (1921–68).

1
BASICS

1.0

The clarity of your thoughts determines the success of your effort.

But clarity is not enough. Your message must get your reader *involved*. (sec. 1.2.4).

1.1

Canadian Marshall McLuhan created a stir in the advertising world with his slogan, "The medium is the message." I disagree. The medium is *not* the message, any more than you are the vehicle that brought you here today. The medium is a *delivery system*, a conveyance, a carrier (the Internet, telephone, radio, letter, newspaper, smoke signal, or carrier pigeon). What goes over or through those is the *message*: sounds, words, symbols, or colors that comprise a *thought*; a notion the originator wants the receiver to know, to use, to share, or to pass on.

An idea is worthless locked into the mind of its generator. To be of any value to society, its originator must transmit it to other people, communicating by words, sounds, or gestures. In this volume I deal only with words, since nonverbal communication is a wholly different subject outside the scope of this present discussion.

Technical and administrative communications are facets of human communication; thus the *objective* of this entire volume is to prove that *a communication must convey a thought in its entirety and unchanged from one mind to another.*

1.2

A while ago I gave a talk to the annual meeting of the American Institute of Certified Public Accountants. Before even a "Good morning," I started with, "Would everyone please raise both hands." Surprised, the audience, still for a moment, began to do just that. Soon enough, when all hands were up, I bade them to relax. I went on to tell them they had just helped me demonstrate four basic notions about communication (sec. 1.2.1–1.2.4).

1.2.1

First: The use of an *amenity*. Remember I asked *please*. It may have blown by in the shock of the request, but it did register on their subconscious, and without knowing why, they did comply. An amenity included in an instruction, a request, or an order (even though they "have to," they feel better about it) helps to penetrate your target's barrier of receiver resistance (sec. 1.7.1), making him a little less unwilling to receive your message.

1.2.2

Second: Word use level. Your communication should be worded to your receiver's comprehension level, generally based on his education. Don't flaunt your erudition to a laborer, nor patronize a nanotechnologist. Be *simple* (easy to comprehend), not *simplistic* (couched in head-patting terms). In ordinary English, I asked that CPA audience, "Please raise both hands." Had I said, "Stick 'em up," some may have, but most would not, while questioning my sanity. But there is no difference in sense of meaning between those two requests. The difference is in the level of the wording (sec. 1.7.2).

If I were to have walked into that room (1) saying absolutely nothing, but brandishing a loaded and cocked Uzi, it wouldn't have taken long for everyone to raise both hands. The Uzi would have done the communicating for my silent armed self (a communication without words).

If I had entered that room (2) packing that same loaded Uzi and loudly demanded, "Stick 'em up!" maybe that might have gotten some attention a bit sooner, but the audience would have raised both hands.

Had I entered (3) with both my hands at my sides, unarmed as I was, and shouted, "Stick 'em up," the audience might have snickered, perhaps a bit confused, but no one would have lifted a finger.

The fourth instance, I've already described: I asked, "Please raise both hands," and they all did.

The point to that exercise is, in the first two instances with or without the message, clearly the Uzi would evoke the positive reaction. To verify, the third instance was the same as the second, but without the Uzi, and that *message* would provoke no compliance. But when I changed the wording of the message to the group comprehension level, it did get the result, without the persuader.

1.2.3

Third: The hands bit demonstrated a complete communication. I had an idea in mind, *to see the audience seated, with both hands up*. I transmitted that idea orally, that was all I could do. The rest was up to the receivers. First, they had to be willing to accept the message. Once willing, they accepted, then received the message. Once received and read (or heard), they understood it. Thus understanding the message, they reacted positively, and up came the hands.

There does *not* have to be a verbal (spoken or written; sec. 1.8.2) response to a communication, if the reaction is a positive response to the original message. Seeing the audience with hands up—my original thought—made it a complete communication. QED (*quod erat demonstrandum*, which was to be demonstrated).

1.2.4

Fourth: By raising their hands, they were all *participating* in that presentation; ergo, they were *involved* (sec. 1.0).

1.3 CONCEPTS: THE SKELETON OF A MESSAGE

You could regard any communication as a unique, three-dimensional, cubic entity, with direction and aspiration. If the cube were a physical thing, by universal convention its dimensions would be length, width, and height. You might wish to call them hither, thither, and yon, or whatever else you'd like, but every one concerned *must* know in advance what your names *mean*, so all are simultaneously at the same page and paragraph. Figure 1.3 gives you the key to what all these dimensions are called, what they mean, and how they all fit together.

Consider standing in a cubic room (fig. 1.3), looking at the north wall, Logical Basis (sec. 1.4); the east wall, Material Substance (sec. 1.6) is on your right; and the floor beneath you is dimension three, Minimum Standard of Acceptability (sec. 1.7). Each dimension consists of several elements.

1.4 LOGICAL BASIS (WHY WRITE) (SEC. 2.1.2)

Logical basis is the first of the three dimensions of a complete communication. *Before* you contemplate complying with a writing assignment, if the

FIGURE 1.3. MULTIPLE DIMENSIONS OF A MESSAGE

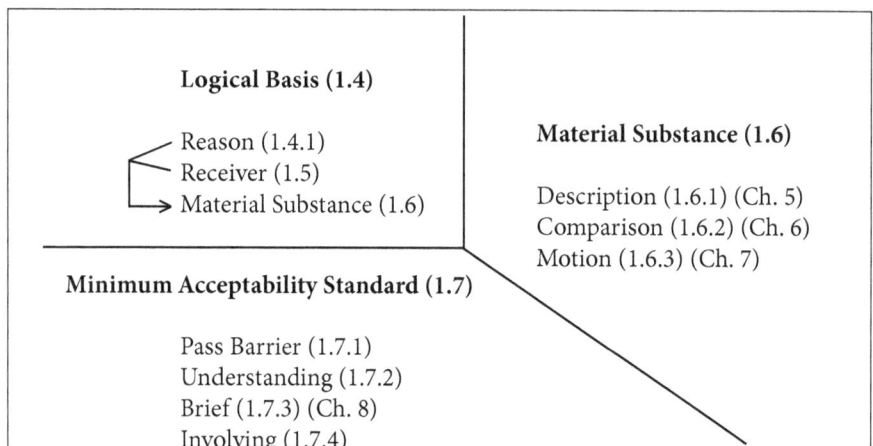

first thing you think of is anything other than asking yourself, *why am I writing this?* you are starting without the foundation of a logical basis. And the answer is *not* "If I don't write it, I'll get fired" or "lose a client." Those answers are *motivations*, not reasons for writing.

From my first teaching year's student work, I took the 200 odd individual reasons they offered and sorted them into generic groups of similar or like nature. Three rounds of re-sorting yielded the same six generic groups. The general nature or sense of direction of each group yielded its category name. As the years went on, student reasons for writing never yielded a single one that could not fit into one of those six categories. Ergo, experience showed the answer to "why write?" is any variation of one or more of these six categories.

1.4.1 TO INFORM

To tell your reader as much as required about the subject of his inquiry, or to inform in general as in a published article. The difference between information and instruction is, information tells *what ought to be done*, instruction tells *how to do it* (sec. 2.1.21).

1.4.2 TO INSTRUCT

To tell *how* to accomplish what needs doing (sec. 2.1.22).

1.4.3 TO INFLUENCE

To try to steer a reader's thinking in the specific direction *you* would prefer. To enhance the importance of some entity. This may be the only way to get your opinion heard when writing *up*, as "You might wish to consider . . ."

A proposal is successful, and you win the job, *only* if you *influence* the receiver sufficiently to agree that yours is the best among the other competitors. Compose your proposals carefully (sec. 2.1.23).

1.4.4 TO CONTROL

When you are an information source, or in a position of authority writing *down* (always add a "please"). To make a necessary or valuable point. *Never* write anything in anger (sec. 2.1.24).

1.4.5 TO CRITICIZE

Before you do, you should be in a position to be able to criticize, then be able to defend your stance. I lay out a complete discussion on the writing and use of criticism in chapter 17 (sec. 2.1.25).

1.4.6 TO RECORD

You should copy your finished document to the archive, and for your own files, especially if it is a group effort and there is a dissenting opinion within. Write for the record if you want to block something. If your writing is a criticism, it should go on record for both attribution and responsibility. Those *copies* should go to authority. Also, to claim yourself as that idea's originator (sec. 2.1.26).

1.5 RECEIVER (TO WHOM)

The second principal matter (sec. 2.1.3), which along with *why write?* combines to determine their dependent function, *material substance*. Writing a proposal, a report, a letter (chap. 18), or a paper for presentation (chap. 13) is different from writing an article (chap. 15) or a book (chap. 16). The most significant departure in the later groups are, in effect, addressed "to whom it may concern," any reader, while you write items in the earlier group targeted to the specific interests of a specific audience.

To decide the language of the message to your target reader, you must add into the mix the considerations mentioned in section 1.5.2. There are no two projects that are "cookie cutter" identical; each job has a life of its own and is different from another, as delineated in the next six categories of receivers.

1.5.1 THE APPARENT AUDIENCE

The apparent audience (sec. 2.1.31) is the person whose name *appears* as the addressee of record (the dean, the district manager, the commanding officer). He is, per protocol, the de facto recipient of correspondence on behalf of the organization. He might not really read your missive; he may delegate the reading, action, and response and may never actually see it. In most cases he is not your target audience.

1.5.2 THE REAL AUDIENCE

The real audience (sec. 2.1.32) is your actual targeted receiver, the one *to whom* you are writing. He is the one the apparent addressee has delegated to read, act upon, and respond to your message, if it is an organizational situation. If he is an individual, then he is at once both the apparent and real audience (sec 1.5.3).

To achieve the maximum results from your message, you must learn as much as you can about your real reader: Who *is* he? What is his background and his education? What are his current surroundings? What is his focused point of attention? And, what is his *knowledge datum*? You must consider the answers to all those questions to form your message. If you write *too high*, he may not comprehend completely, become frustrated, and either ask someone else (who you are not addressing) or just pitch it. If you write *too low*, before your message gets up to his knowledge datum, he'll become impatient and pitch it. You need to determine at the outset what he needs to know, and what he wants to know from you.

1.5.3 THE INDIVIDUAL AUDIENCE

The individual audience is your real reader; thus every detail discussed in section 1.5.2 applies (sec. 2.1.33).

1.5.4 MULTIPLE AUDIENCES (SEC. 2.1.34)

This section is *not* about a message addressed to an individual reader (sec. 1.5.3), who then passes it along to one or more additional people (sec. 1.5.5). Rather it concerns one single communication addressed to a connected receiving group acting as a unit on the same level. I can illustrate that point best by detailing this situation:

As a principal in an engineered-construction firm, you wrote a proposal in response to a request for proposal (RFP) from a small city to design and construct a wastewater treatment plant as a turnkey job. Although not the lowest-priced bidder, you won because you were resourceful, then strategically used good judgment with the information you found. All the losers submitted one-size-fits-all bids that never "reached" the adjudicating body, the city council.

That group (fig. 1.5.4) consisted of the mayor (M), whose main concern was the effect that the project would have on his image at the ballot box come November; the city comptroller ($), whose interest was in seeing how much the city could get for the least money spent; the city attorney (L), whose interest was in seeing how much the city could get for *no* money spent; the city health officer (MD), whose genuine preoccupation was the people's good health; the city engineer (PE), who cared about supervising the construction, getting the plant on line, and learning the operation and maintenance of it after the contractor left; the city communications director (PR), who worried about how many inches of ink this project would get in the paper at the state capital; and the citizen taxpayer (c), who wondered what benefit he would gain for his increased tax dollars.

Looking at the council's constituency (fig. 1.5.4) and considering each member's concerns, you realize that a simple proposal detailing every necessary aspect would become a horrendous task of wasted effort because not every member will digest the entire text, nor its implications. Perhaps because of differing interpretations it might even be counterproductive (as it was with the losers). Instead, you evaluated the concerns of each and thought the ones most responsibly involved were the PE and MD; thus you wisely decided to focus your bid on those two, without so stating in the text: don't let anyone feel ignored. You added as one appendix a detailed full spreadsheet with notes expressly expanded for $, since that was all he really needed and wouldn't be involved in the details of anything else. Your (correct) thinking was that no single proposal would gratify everyone, so if you focused on two individuals (MD and PE) and thoroughly informed $ about his interest, the rest would go along. They did.

FIGURE 1.5.4. SINGLE COMMUNICATION TO MULTIPLE AUDIENCES

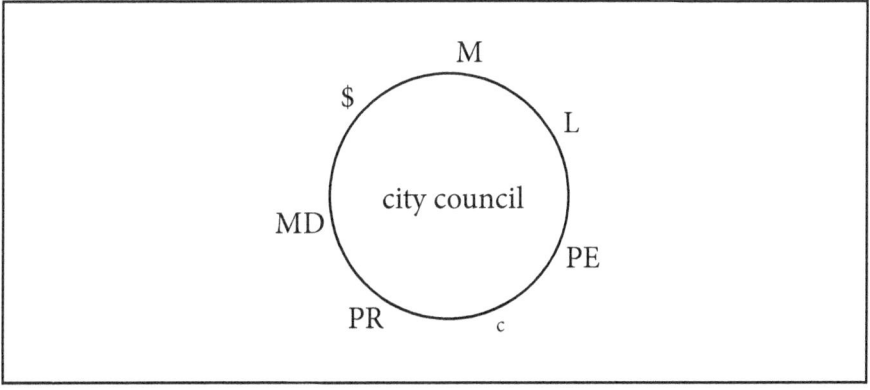

The same procedure can be applied to a graduate student faced with presenting his thesis to his faculty committee. He won't please them all, but by focusing, he *can* please his champion.

1.5.5 PRIMARY-LEVEL AUDIENCE (SEC. 2.1.35)

This is the antithesis of the multiple audience discussed in section 1.5.4. This section deals with one message on a single subject to several people who may be on different levels. You focus it to the comprehension and knowledge levels of the primary person in an ad hoc group of various levels, all of whom get copies. You may add an appendix, expanded with details, for another member of the group, but on the same level as the report. Thus the message is targeted to the primary receiver, and the rest have to read it (up or down), and the receiver replies on behalf of the group. For instance, an executive of an electronics equipment manufacturer generates an idea for a new product he thinks will do well. He issues a request to his chief engineer to ascertain if it can be fabricated in-house. He sends copies to budgeting, to advertising, and to marketing, but the message is on the engineer's level. The others have to read it (up or down) and hold it as a heads-up for later contact, and for input to the primary for his reply.

1.5.6 SUBORDINATE LEVELS

Subordinate levels are the recipients of the primary message (sec. 1.5.5), not necessarily on the primary recipient's level, but understandable to them. The original message also may contain some material for or about any one on the list other than the primary receiver. In effect it is as if they were reading the message over the shoulder of the primary receiver (sec. 2.1.36).

1.6 MATERIAL SUBSTANCE

Depending upon why you write (sec. 1.4), and to whom (sec. 1.5), your finished product, the meat, your *message*, is comprised of three elements: descriptions, comparisons, and motion. If you compose them skillfully, they will convey your ideas to your reader in his most easily understood form.

1.6.1 DESCRIPTIONS

Descriptions tell what things *are*. From them your reader should be able to envision precisely every entity you relate.

A detailed discussion on the composition and use of *description* is laid out in chapter 5.

BASICS 9

1.6.2 COMPARISONS

Comparisons tell what things are *like* or what they may be worth: the relative merits of two or more entities with regard to each other, or with regard to a given standard. You will find a detailed discussion on the composition and use of *comparisons* laid out in chapter 6.

1.6.3 MOTION

Motion (including instructions) tells how things *move*. Descriptions and comparisons are *static*. There is a detailed discussion on the composition and use of *motion* in chapter 7.

1.7 MINIMUM STANDARD OF ACCEPTABILITY

You can achieve only *half* a complete communication. All *you* can do is construct a message and issue it via a medium. It is then out of your hands. For example, a quarterback heaves a pass to a potential receiver. Downfield his tight end wants to catch it, he does, then he knows to tuck it in and run with it. Similarly, to complete your communication, your target reader must *want* to receive your message, read it, understand it, and react to it. *The degree of receiver reaction determines the success of your message.* There are several things you must do to achieve your goal.

1.7.1 PASS HIS BARRIER OF RECEIVER RESISTANCE

Today, everyone has barriers up to shield himself from the ceaseless barrage of unsolicited material, spam, and junk mail. Herein lies the philosophical conflict built into human relations. Simply said, everyone in our society guards his right of privacy, and also enjoys his freedom of expression. For you to exercise your right of free speech, you must infringe upon your receiver's right of privacy. And for him to maintain his right of privacy, he curtails your freedom of speech. The resolution, of course, is in compromise and reasonability. You must attract your potential receiver with an amenity, or with an interesting title, headline, or lead line, or tell him early on what he gains from reading your complete message.

1.7.2 BE UNDERSTOOD

Once past your reader's resistance barrier, your now-willing reader must understand what he reads, to be able to put your message to mutual advantage. Addressing him on both his proper comprehension and knowledge levels (sec. 1.2.2) is the only way to write: too complex or too simplistic and you lose him.

1.7.3 BE BRIEF

No one enjoys reading voluminous copy unless he needs to, or he is immersed in the subject. Make your communication *commensurate in size* with the *value of the message* it carries (chap. 11). You cannot always produce a document as tight as you would like, but stay with the information you want to convey; avoid including abundant peripheral or unrelated material. In chapter 8 I lay out a method for shrinking your communication while still retaining all relevant substance. But take care you don't brief your message out of existence.

1.7.4 GET YOUR READER INVOLVED

Assuming your target *does* read all you've written, then shrugs, puts it aside, and dismisses it from his mind, you have lost your point. Worse, you've wasted his time, and yours. By reading your effort, he is "hooked." Now your document must reel him in and cause him to go along with whatever your message seeks. Your improved writing style (chap. 3) helps you to follow fast after passing his barrier. Once you have his attention, expand on your theme. Just like that, you have him involved.

1.8 DIRECTION

A complete communication follows this critical path (fig. 1.8): origin, transmittal, receipt, response.

1.8.1 ORIGIN

You conceive an idea, or find new information, or respond to another's message. This is the notion you have in mind. To be useful it has to get out.

1.8.2 TRANSMITTAL

Verbal means *the use of words, both spoken and written*. Thus you compose your message according to all the suggestions in this book. Then you issue it, *vocally* or in writing, via your choice of medium. That is all you can do: generate a notion, organize it, and send it off. The rest is up to your receiver.

1.8.3 RECEIPT

Your intended receiver must be willing to accept your message or it goes nowhere. Or, by consent, he does receive it, reads it, understands it, then reacts.

FIGURE 1.8. PATH OF A COMPLETE COMMUNICATION

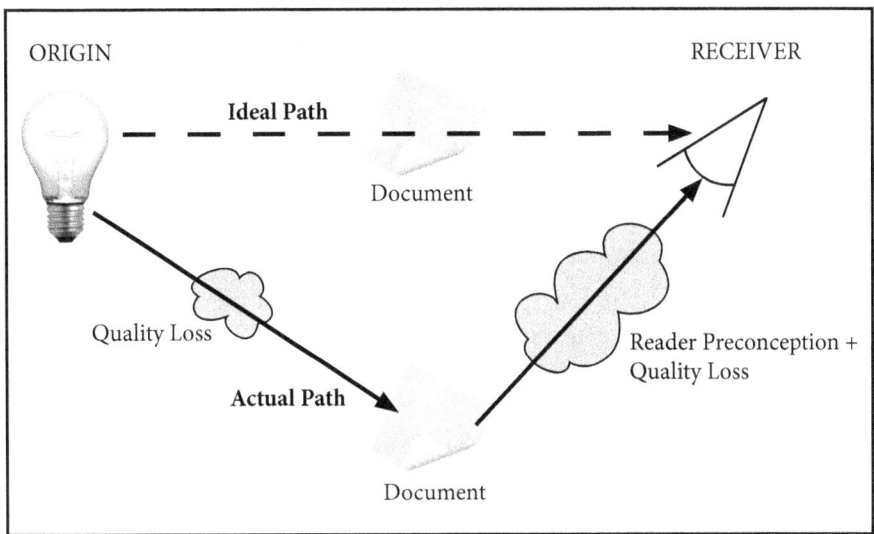

1.8.4 RESPONSE

His reaction to your message determines his reply, and the success of your message. It could be yea, nay, or tabled. His reply does not necessarily have to be verbal: an *action* could suffice. But by that responsive reaction, he makes the transmittal of your idea a complete communication.

1.9 CLARITY

Refer again to Figure 1.8. The incandescent light bulb (upper left) is symbolic of the human mind. When the mind generates an idea, the light goes on. To be of value, the idea must be transmitted somewhere. When the notion of specific gravity tumbled to the bathing Archimedes, allegedly he ran wet and naked through the streets of ancient Syracuse shouting, "Eureka, I've found it!" Today the constabulary takes a dim view of streaking. Besides, now we have many more ways beyond the town crier to broadcast news. The printed page (lower center) is one. The willing reader (upper right) ingests the message. Theoretically the receiver should then (dashed line) have the same mental image as the sender, but most often he does not.

Between the sender's mind and what he commits to the paper, there naturally occurs a modicum of loss of quality, a bit of clouding. Thus the printed record of the image is slightly less acute than the sender's original thought.

No person is devoid of feelings. That built-in prejudice, however small, has some subconscious effect on receptive thinking. That in turn causes more clouding and more loss of quality from paper to mind. This clouding, added to the clouding from the originator to the paper, leaves a diminished transmittal. Your task is to try to make the thought you send as nearly perfect and as carefully crafted as your talent and skill allow, then the loss of message quality is all on the reader.

People don't read what they see, they read what they think they would like to see (the clouding from paper to the reader's mind). That is the principal cause of receiver message clouding. Long ago I was driving from here to there with my then ten-year-old son. We stopped for a red light beside an auto supply store. He saw a store window sign and asked me, "Papa, why is *that* store selling *muffins and pies*?" I looked, and told him to read it again. He did. The sign appropriately offered *mufflers and pipes*. It turned out that he was hungry; that subconsciously became foremost in his thinking, clouding the transmitted idea.

Another time I was an expert witness for a lawyer in a traffic case. He harbored an unfortunate deep dislike for traffic policemen. In the ready-room he picked up a stray copy of a newsletter and seriously said, "Look at them, now they even have a publication, *The Retarded Patrolman*." Scanning what he handed me, I was moved to say, "Come on, Jake, that's *The Retired Patrolman*."

An invitation to a dinner meeting offered a choice among three entrée selections; one was "Half roasted chicken." Who would eat semiraw meat? Or, if they meant, "Roasted half chicken," was the other half badger? The confusion vanishes simply by the addition of an *a*, "Half *a* roasted chicken."

Those are preconceived receiver notions that must be overcome. Take the time to write what you *mean*, what you intend to be read. If you think them through and organize your ideas carefully, your message will be transmitted clearly via your chosen medium to your reader. The clearer your thoughts, the more transparent your message, and the more successful it will be.

> . . . COMMUNICATION, DRAWS ALL, EVEN THE MOST BARBARIAN NATIONS, INTO CIVILIZATION.[1]

NOTE

1. Karl Marx and Friedrich Engels, *The Communist Manifesto*, trans. Samuel Moore (New York: Penguin 1967; orig. pub. German 1848, English 1888).

2
ORGANIZATION AND DEVELOPMENT

2.0 ORGANIZATION

Organization is a preplanned, orderly, methodical procedure to make an army out of rabble, or a football team from a flock of fumbling freshmen. First, however, you must have an idea, a theme, or a central thought line.

2.0.1 PREMISE

Any activity can be regarded as a data-gathering phase to acquire substantive information for a subsequent written effort. Conversely, any written effort can be regarded as the recorded documentation (or the documented record) of its foregoing activity. Thus, we are looking at the two like ends of the same municipal ferry boat.

2.0.2 APPLICATIONS

You can use all the theory from chapter 1 in *actual practice* right now. In performing these activities you should (1) learn the ways to *apply* the theory; (2) follow the construction of a prototype project from assignment to deadline; and (3) build your own draft document from your own assignment to deadline. Thus, by the end of this chapter, you should have constructed a rough first draft as a result of your effort.

First you must select a project for yourself to document, as if someone had assigned it to you. It is best if you take it from your job at hand or something else real and useful. A real assignment is best.

Figure 2.0 is a schematic teaching aid that delineates all the steps and points in this procedure, and shows their interrelationship from start to product. All that will be detailed later in this chapter.

Begin at the bottom with the horizontal assignment (A) line (fig. 2.0). Some reasonable space above A is your deadline (D/L) (fig. 2.0). The space between A and D/L, time (T), is divided into five strata, or stages—the development periods. There are always five of those, and always in the same order. Although shown in Figure 2.0 as being equal (1/5, or 20%), *they never are*, since every project has its own life. The proportions may be different sizes but must total to 100%.

The stages, in order, are Thinking and Planning (P) (sec. 2.1); Research and Investigation (R) (sec. 2.2); Data Reduction (D) (sec. 2.3), blending (through a gray area) into Outline Building (O) (secs. 2.4–2.6), and Writing (W) (chaps. 3, 4, and 9). On the horizontal between P and R is focal point f, and directly above it on the line between O and W is the dispersion point d. All your findings in P are represented by root lines converging at focal point f. The communication grows as a single entity between f and d when it multifircates into whatever the written product becomes (chap. 9).

2.0.3 THE PROTOTYPE PROBLEM

The Federal Building at Foley Square in New York City is a forty-story bronze and glass tower personifying the U.S. government. It was built and is operated and maintained by the General Services Administration (GSA).

You are the building manager (BM), with the comparable status of a ship's captain. You arrive at your office one morning and your computer

FIGURE 2.0.   ORGANIZATIONAL PROCEDURE

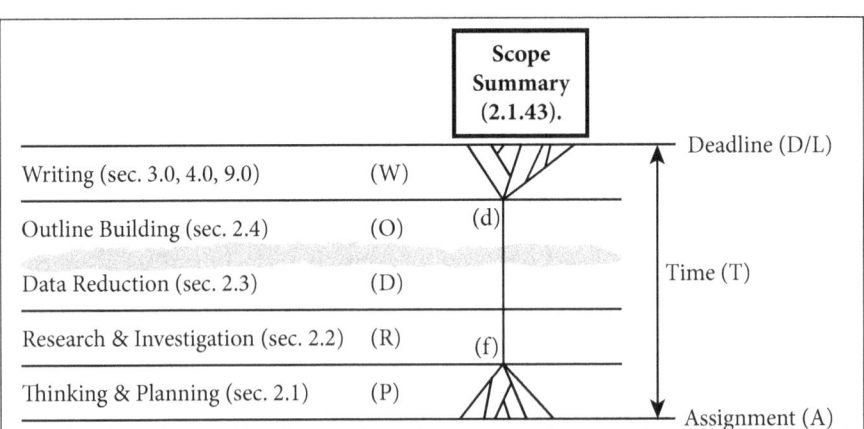

greets you with this e-mail message over the signature of the "GSA Coordinator of Buildings":

> *Report to this office within three weeks on the economic feasibility of operating a chicken farm on the roof of the Federal Building in NYC.*

Unhinged, you request to be relieved of the assignment because you know nothing about chicken farming. GSA responds immediately:

> *You are not reading the assignment correctly. It is not about chicken farming, it is about economics. BM, surely you know about economics?*

You must read and *understand* every assignment thoroughly to be able to *undertake* it thoroughly. You then go to human resources in your building for help. Enter your consultant, I. You and I begin with Figure 2.0.

Figure 2.0 is a schematic guide to the text in the rest of this chapter. I designed it to depict and clarify the steps in building a document from assignment (A) to completion at deadline (D/L). The "communication growth tree" starts with the root lines you develop in the Thinking and Planning stage (P) (sec. 2.1). All the roots converge at focal point f on the line between P and the Research and Investigation stage (R) (sec. 2.2), where the "tree" grows up through the Data Reduction (D) (sec. 2.3) and Outline Building (O) stages (sec. 2.4) to dispersion point d, where it branches out into whatever the written product (chap. 3) becomes, to respond to A.

I repeat, although for clarity I show the five stages as equal sized (1/5, or 20% each), actually they hardly ever are. Every individual project is unique unto itself. A doctoral candidate in geotechnical engineering writing a thesis on foundations in varved clays may develop his P in two weeks; spend his academic year in the laboratory generating data, stage R; do his D and O stages in a month; and spend three months writing, W.

At the other extreme, a magazine writer may spend days developing an article, research it in two weeks, do an outline overnight, and write it all the next day. The diagram equality just makes the text easier to envision.

2.1 THINKING AND PLANNING (P)

Where to start, and how to go. Most engineering students were told in English 101 to start a report by confecting an outline, naturally from within the ken of their own knowledge; beef it out with researched information; and write. That *may have been* the right way then, but not *now*. That outmoded approach does *not* work realistically in today's working world. We can't eat archaic and still have it. You can't tell what you may have missed because you don't know what you don't know.

2.1.1 VIABILITY

Start by ascertaining if the *whole project*, not just your report, will actually work. If the holistic effort proves unviable, your report will be also. Don't waste your time.

In this case you check with the EPA people, who determine there is nothing in the project inimical to the vicinity's environment. Your *local* canvass shows a promising market for your product. Finally, the building's engineering design data reveal that the miniscule additional load poses no further strain on the total structure. Ergo, finding no *negating features,* you can say the project *is* viable, as will be your report; so proceed.

2.1.2 REASON FOR WRITING (FIG. 1.3, SEC. 1.4)

The reason(s) selected (sec. 2.1.21–2.1.26) predetermine the *use* your receiver will make of your effort, which, in turn, determines its useful longevity. You can write to inform, instruct, influence, control, criticize, or record.

2.1.21 INFORM

Yes (sec. 1.4.1). In effect, A says, "Give me information about . . . ," and you want to do that.

2.1.22 INSTRUCT

No (sec. 1.4.2). Think: If GSA wanted instructions for *operating* the chicken farm, A would have included "and give instructions for its operation." Go the extra mile, yes, but *don't* undertake big items *not* requested in A; you'll add 150% unnecessary work for yourself. Also, *if* GSA *wanted* such instructions, they would have gone to Cornell University's Department of Poultry Science, not to a BM versed in other areas.

2.1.23 INFLUENCE

Depends (sec. 1.4.3). Unless A specifically asks for an opinion, don't offer it. GSA requested facts, yet if the preponderance of evidence shows clearly it is a bad bet, you should *suggest* so.

2.1.24 CONTROL

No (sec. 1.4.4). You are not in a position to control what goes on at GSA headquarters.

2.1.25 CRITICIZE

No (sec. 1.4.5). But if you harbor strong feelings based on the evidence at hand, you might try (sec. 2.1.23) subtly to influence (sec. 1.4.6).

2.1.26 TO RECORD

Yes. You want to go on record as having been responsive to A. Or if you should find some additional aspect, take credit and *responsibility* for it.

Now you have information, possibly influence, and to record, as root lines in P, your reasons for writing (sec. 2.1.2).

2.1.3 RECEIVER(S)

Section 1.5 precisely details your receiver levels, but you cannot take them at face value. Since you are able, do some inquiring to learn more about the target individuals. You may not always be able to do that, but it is worth the effort to try because each item discussed in this section is another root line in P (sec. 2.1).

2.1.31 APPARENT AUDIENCE (SEC. 1.5.1)

Since the GSA coordinator of buildings signature *appears* on A, *apparently* he becomes your *addressee of record*, but you have no certainty that he was the originator, or will be the only one to act on it; and there is only a 50% chance he may even see it. He is *not* your target.

2.1.32 REAL READER (SEC. 1.5.2)

Probing at GSA reveals that the signatory of A (the apparent reader) has three deputies. The third, born and raised on Maryland's Eastern Shore, knows nearly everything about chicken farming and the economics thereof. He was the originator of this notion and will adjudicate your report. He is your *real* audience. Your report must answer to both his level and to his needs and wants, all discussed in section 1.5.2.

2.1.33 INDIVIDUAL (SEC. 1.5.3)

Deputy three, your real audience, is an individual, as detailed in section 1.5.2 and 1.5.3.

2.1.34 MULTIPLE READERS (SEC. 1.5.4)

This does not apply in this project.

2.1.35 PRIMARY READER (SEC. 1.5.5)

Although your report may pass through many hands, the focus must be on your *real* reader (sec. 2.1.32) with appendices targeted to *specific* secondary readers, *if* you need them.

2.1.36 SUBORDINATE (SECONDARY) READERS (SEC.1.5.6)

Your inquiries at GSA reveal that the budget director is thoroughly involved in the fiscal aspects of this project, but has only marginal interest in the rest. To satisfy that requirement, you deal with the money section as with the other report sections, then you *add* a thoroughly detailed appendix addressed to the budget director as your *primary* receiver for that *specific* appendix only.

2.1.37

Background, education, position, and ability considerations for your *real* reader *only* become your next thoughts in determining your writing levels. Focusing your report on the real audience is essential; all others have to read as best they can. There is no strength in a one-size-fits-all report.

2.1.38

Surrounding conditions of the real reader, climate, weather, budget, politics, policy, personnel, and personal matters all exert some form of constraint or have an influence on your real reader's reaction to your report. Look into those considerations as best you can before you start. They are all root lines in P.

2.1.39

Knowing your real reader's knowledge datum will help you decide at the outset what he *needs* to know and what he *wants* to know from you. Consider all the foregoing material as you build your report.

2.1.4 MATERIAL SUBSTANCE (SEC. 1.3)

If I were reading this report instead of writing it, what would I want it to tell me? That is the question you must ask yourself now to enable a proper response to the charge in A. You have already determined *why* you are writing, and *to whom*, the two basic elements. Use both those factors to construct *what* you tell the *real* reader.

2.1.41 SUPERFICIAL PRERESEARCH

Attempting to make this report as useful as the real reader would like it, you approach the task from his position. You list as many questions as you think he might ask you if you were to interview him personally. Of course, that does happen on rare occasions, thus now you *act* in his behalf. There are times when your own knowledge of the subject is thin, thus you are actually asking questions in your *own*, and his, behalf.

There are two extremes to listing questions: First, you may be so steeped in the subject that you can't remember when you didn't know all there was to know about it. Then it becomes difficult to choose just how far back you go to reach level land without being simplistic or patronizing. You just need to use good judgment, or ask someone who knows, but less than you. The other extreme is when you know so little about a subject you don't even know what *kinds* of questions to ask. This is the point where you do *basic* research, not to gain specific detailed factual knowledge for the report content, but to acquire a sufficient patina of information to enable you to ask pertinent questions.

This is also the first of two chances (sec. 2.1.422) where you can find material you had not previously realized about the subject. Now you have enough basic facts to generate intelligent questions and start to build a preliminary outline.

2.1.42 PRELIMINARY (PROVISIONAL) OUTLINE

This is the point where you start building the preliminary structural skeleton of your document. This is also where you begin to see the things you didn't know you didn't know. Begin with a blank screen, or a blank pad of lined paper. Dismiss most other things from your mind. Then, think under the hat of your real reader.

2.1.421 QUESTIONS

As if you were he, inquire about what he'd need and what he'd want to know concerning the economics of operating a New York City rooftop chicken farm. As his surrogate, you start by compiling a list of random questions, one per line, no boundaries, no order, and maybe include a few as from subordinate readers. List the questions as they come to mind. Being a "city fellow," there are plenty questions you could ask.

If you think you might have asked the same question earlier on the list, don't go back through the list to check; that breaks your forward momentum. Just add it and move on; you'll have a chance to pick out those redundancies after the next step. Don't dawdle, because your project clock is ticking, but don't be too hasty, because this is the seeding stage for further growth. Carefully use good sense. Too few questions will result in gaps in the text; or you could choke to death on too many. Here, you have stopped at sixty-one listed questions.

2.1.422 PRIMARY GENERIC SORT

Were I to dump a pile of miscellaneous small hardware on your desk and tell you to "do something useful with it,"[1] most probably you would sort the pieces by piling washers with washers, screws with screws, wing nuts with wing nuts, and so on until the original pile was dispersed into several smaller piles of items of *similar or like nature*. You do the same routine now with your pile of questions. You sort that list into smaller lists of questions of similar or like content.

N.B. (*not* a law, but a useful rule of thumb): *If you have about twenty-five questions, they would sort into two or three groups; about seventy-five questions would create five to eight groups of similar subtopic questions.* Since you have sixty-one questions, you set out eight blank sheets, allowing for the maximum sort. Here is where you catch those repetitions from the original compilation; keep the better and toss the other. Should you run onto a question that appears to fit into more than one group, put it into each list for now. You'll catch it later. Be sure all your questions are sorted. And, *no* catchall group. Your sixty-one have fallen into seven lists of similarly natured questions. Return the eighth, unused sheet to stock. Do not yet rearrange the questions in any group.

Look over each list individually. Switch any one question if it seems to fit better in another group.

The dust having settled, you now have seven condensed sheets of *primary* sorted questions for a new combined total of fifty-two questions. But two group sheets have only one question each. There is nothing wrong with having a one-question group, if indeed that one question is *unique*. A closer look shows that one of them would work in a larger group, and once moved into there, it seems to connote more material than an entry already there. Eliminate the weaker one, add the one from the solo group, and you have six groups. Repeating the process on the other single-entry sheet, it can be shoe-horned into a different, larger group, but there is a stronger item already there. You delete that question, thus the second single-item list is gone. Finally there are five groups; that fits the minimum recommended by the rule of thumb.

2.1.423 GROUP HEADINGS, PRINCIPAL MATTERS

Your project has now shaped to the point where you cease dealing with it holistically. You have five distinct lists of questions in no particular order. Within those collective lists lies the nucleus of your document. Your task now is to dig it out and process it into a finished, polished, presentable product.

Deal with each of the five sheets one at a time. On any one sheet is a list of five or six questions related by a common idea. As you reread that list, the common thought will float to the top, thus the list literally will name itself. For this batch the group names become Chickens, Plant, Funding, Marketing, and Operations, not necessarily in that order, at f. Those five headings are for your report's main sections, or chapters; they are the *principal matters* for your contents and your keywords for information retrieval. All five of them must appear in your abstract (sec. 8.4.2). Do not yet arrange them into a fixed order, nor do you yet organize the questions within each group.

In several earlier paragraphs I cautioned that creating an outline from your own knowledge eliminates all unknown aspects. The constraint that you don't know what you don't know is liberated by the questions route. Being a city-bred person, you had a mental image of fowl at large in a barnyard. You had not the faintest idea of a *planned, commercial plant*. That discrepancy would have lessened your report, which would have required later serious patching. By taking the questions route, natural queries led to the revelation, and inclusion of the previously unknown but necessary section on Plant.

2.1.43 SCOPE SUMMARY

If you use the names of your not-yet-organized groups, add your reasons and your real reader, then you have the basic ingredients for building an outline, but you need a destination indicator. Just as the pin flag on the green shows the golfer at the tee 300 yards away where to drive his ball to the hole, *you* should have a target marker.

Your Scope Summary (SS) is that target guide. It is an aspiration, an end-result specification, a memorandum to yourself detailing where you want to be with your report when you are finished, and what you want it to tell. All those things become one in your SS. This is *different* from your abstract (sec. 8.4). That can only be written *after* your report is finished. Now, *you* write your SS *before* you go another step. Write the SS on a file card (fig. 2.1.43) so it will last the full project time. In 50± words you should be able to tell yourself what you hope your product will be. Include some form of the word(s) that indicate your reason(s) for writing. Figure 2.1.43 shows two excellent SSs, one from industry and one from government. The end products of both were equally superior and useful to their jobs.

FIGURE 2.1.43. SAMPLE SCOPE SUMMARY CARDS

> **Scope Summary**
> The APICS USERS GUIDE will *inform* plant supervisors of correct user procedure. It will also serve as a reference, and an *instruction* manual for problems arising from operator error and/or system malfunction.
>
> John Bradberry
> American Can Company

> **Scope Summary**
> This report will *describe* the infrared imaging-and-tracking system (IITS) and its use. Text will include the reasoning for the design, implementation, and data obtainable. Also, it will cover the *procedures* for installation, calibration, operation, and maintenance.
>
> Laird Moffett
> U.S. Naval Res. Lab, D.D.

Your SS is your navigating guide during the entire project, yet it never appears per se in the document. The sole exception to that rule is: If you write the SS skillfully, then either as it stands, or if you change one or two words, it can become an excellent opening (introductory) paragraph for the document, written in the *present* tense, of course.

With that in mind, take another look at Figure 2.1.43 to see how right that notion can be. It may spare you some anguish when you actually write, at W.

2.1.44 PASS KEY TO BUSY PROSPECTS

Here is another positive use for your SS skill (secs. 13.5.1 and 13.5.2). Early in my career I was a sales engineer for a portable steam generator manufacturer. I applied the generators to be mounted on the counterweight to make pile-driving rigs self-contained. It was difficult as a "stranger" to get in to explain that innovation to top people or purchasing agents at construction firms. On the blank backs of my calling cards, using the SS technique, I hand lettered in fewer than fifty words the benefits of that combination; it was easy for a busy executive to read. Most times that gained me entrance; many times it resulted in sales.

2.2 RESEARCH (INVESTIGATION), R

You could not have gotten this far in college or the professional working world without having learned early on to delve into and use new information from research. Thus, I shall not presume upon your intellect with that rudimentary drill. But I *can* impart several serious suggestions that will make your research, (stage R) in this system more efficient.

When you've gotten all those ingredient root lines (sec. 2.1.43) focused at point f, you have completed your Thinking and Planning stage, P. Now, guided by your SS, your activity "tree" line grows toward dispersion point d.

Plan and proceed with investigating to acquire all available information and data to fill out your outline structure.

2.2.1 STATIC RESEARCH

Pore over catalogs, indices, data banks, Internet search engines, and libraries. Make use of the observations of others.

2.2.2 DYNAMIC RESEARCH

Interview people, and get out into the field to make *your own* observations.

2.2.3 ORIGINAL RESEARCH

When neither static nor dynamic research yields the information you seek, if the point is not vital to the end result, just mention it (so it is not ignored) and move on. If it *is* vital and there is no available material, then make a small similar project of the item in question. Use the results as an inset in the proper place, if it is not too voluminous, or put that package into the appendix. The late author Studs Turkel, in compiling his nonfiction books, said, "Curiosity killed a cat, but I'm no cat!"

2.2.4 ETHICS VERSUS ENTERPRISE

Use whatever clever above-board methods you can conjure to get material, but do not use fallacious or improprietous means.

2.2.5 AVOID "ONE-STOP SHOPPING"

There is no single source, generally, for input material (except in the unique case of a single subject in a dedicated environment). For this chicken farm project, you have interviewed a senior Department of Agriculture county agent who could answer knowledgeably all your questions about chickens and plants perfectly, yet in *far* too much detail. However, he is a poor source for *good* information on funding or marketing. If you can, seek *only* basic root sources (sec. 2.3.3).

2.2.6 PREPARATION FOR RESEARCH (STAGE R)

You now have five lists of questions and your SS. The *answers* to those categorized-but-not-further-organized questions will provide the material substance of your report. Go out and get them, but *first* set yourself a research plan for where to find your static and dynamic answers. Allow yourself a proper proportion of time (T) to get what you need before it eats into the time you will need for D, O, and W to be finished by D/L. Try to set your daily travels into a paternoster routing to eliminate recrossing or backtracking. Do not research one whole list at a time, nor prioritize, nor organize the listed questions now. Try to get as many answers as you can across the board at every stop.

Sometimes you will get a variety of answers to one question. Do not take them all, but use your SS as a guide to which of the choices will fit best into your project's line of march. When you get an answer, cross out (but don't obliterate) the listed question. Do not destroy the used question lists. Put every answer on its *own sheet of paper*. Paper is cheap, confusion isn't. Mark every answer sheet with a *keyword* or *phrase* that best describes what is on that sheet or graphic. En route you will acquire photos, sketches, magazine article tear sheets, computer printouts, and other graphics. Mark them with *keywords* also.

When I devised this system I'd toss every marked answer sheet into a briefcase. When R was finished, I'd waste valuable time playing a post office-type routine sorting the answer sheets into (in this case) their five piles, then writing an inventory list of each pile. Quickly I thought of a better way: Start each R phase with (here five) empty blank 10" × 12" mailing envelopes. In large letters, mark each envelope across its top with the name of one question group list. Then, as you acquire every new sheet or graphic, *marked with a keyword or phrase*, toss it into its appropriate envelope and list it on the envelope face under the correct group name. The envelope front then becomes an *inventory face sheet*. Thus as-you-go instant sorting and inventorying.

My computerwise amanuensis, Elaine Guidero, suggests this computer-based alternative to the envelopes with face sheets:

> For graphics and material in electronic form, design a hierarchy of folders on your computer: a folder for the entire project; then subfolders for each group heading. Those subfolders equate to the envelopes just described.
>
> Place each research-found item into its appropriate subfolder. Add a text file for comments or further information. Annotate each folder's text file with references to the specific entered item or

graphic. The folder with attendant text file serves as a summary of the folder contents; thus as your face sheet.

If your documents are in programs with annotation capabilities, use them. Another way to keep track of comments or of multiple files is to use a citation management program (EndNote®, for instance), or a wiki-based Web site or service.

Acquire as many answers to your listed questions as your resourceful diligence will beget; you won't get them all.

Budget your R time with reasonable judgment (as with everything else in life). You are done with stage R; on to D.

(That system has worked successfully for me in authoring four books and as the editor of ninety-one volumes of Practical Construction Guides for four major publishers, among other items. The American Society of Civil Engineers named me the recipient of its Peurifoy Construction Research Award in 2001 for using it successfully.)

2.3 DATA REDUCTION AND DISTILLATION, D

You are back in your office with five envelopes, replete with gleaned information in response to your five lists of questions. Your *full report content* is in those envelopes, in its *most primordial form*. You will evolve that material substance from the original disarray into organization via these stage D steps:

2.3.1

Working with full envelopes is inconveniently clumsy, in addition to the risk of losing material. Photocopy all five marked envelope fronts to give yourself five manageable inventory face sheets, the lists of all the available material. Immediately store the full envelopes in a safe, cool, dry place. You will need them when you begin to write, at point d. File away the crossed-off lists of questions for later use (sec. 2.3.36 and sec. 10.2). Those five inventory face sheets are actually your complete report. All you need to do now is organize them into proper sequence and fill in researched details.

2.3.2 ESTABLISH PRIMARY SORT MAJOR HEADINGS

Lay out the five inventory face sheets on your desk in the random order in which they flop. Left to right that might be Chickens, Funds, Operations, Marketing, Plant. That sequence does not appear to answer the charge of A nor hit the mark of SS. Shuffle again, maybe in descending order of importance:

Marketing, Funds, Chickens, Operations, Plant. Still no go. Try chronological order: Funds, Plant, Chickens, Operations, Marketing. That arrangement answers A, according to SS, because it is both *chronological* and *logical*.

In dealing with the federal government you need an encumbrance in front, *Funds*. With those Funds, you build a *Plant*, you populate it with *Chickens*, and *Operate* a farm. Then you *Market* the product. If over a given period you can sell for *more* than your Plant amortization and your Operating costs, you show a profit, and the notion is *economically feasible*, your objective.

Number your inventory face sheets with whole Arabic numbers left of the first decimal point (sec. 2.4): 1.0 Funds, 2.0 Plant, 3.0 Chickens, 4.0 Operations, 5.0 Marketing, your primary sort.

2.3.3 SECONDARY AND LESSER SORTS

From your primary sort (sec. 2.3.2), your inventory face sheet layout, *still unsorted*, should look like Figure 2.3.3. The wavy lines simulate the key thought (word or phrase), now ready to be sorted and arranged into an outline. Look at each sheet list with relation to all the rest. Now perhaps some entries need to be consolidated within their own group, or moved to a more relevant group. Use a critical eye; your reasonable judgment and your SS should guide you to the needed amount of available inventoried information to answer the charge in A. As the quantity of items now stands, 1.0 Funds is OK; 2.0 Plant is not: insufficient information. Return to your question sheet (sec. 2.3.1); obviously you were not able to find responses to most of your questions. Go out to get some more answers. In 3.0 Chickens, you have far too much material (sec. 2.2.5) in relation to the other groups. Carefully prune many of the trivial, less-important listed items, your shaping being guided by your SS. 4.0 Operations, like 1.0, is OK, as is 5.0 Marketing. You will find much more about this balancing act in chapter 10. But *wait*, don't jump ahead to it just yet, for now *stay* with this process.

FIGURE 2.3.3. INVENTORY FACE SHEET (UNSORTED)

Because each sheet is a separate report section (chapter) and has a life of its own, you deal with each individually (and not necessarily in rotation). Arrange and develop each sheet in the same way, however. For this project procedure, you choose to begin with 5.0 Marketing.

2.4 THE DECIMAL-NUMERIC INDEXING SYSTEM

Today we live in a digital world, marked entirely by sequences and combinations of Arabic numerals. Time, speed, dates, temperature, blood pressure, global position, etc., are all noted by numbers, as is contemporary technical and administrative writing.

As late as the mid-20th century, most instructors of English composition on all levels taught that outlines are organized in the so-called Harvard system. In that (generally, not always), ideas are indicated in descending order of importance: principal matters, by Roman numerals (and if you do that, why not be consistent and write the text in Latin?); secondary sort, by capital letters; tertiary, by Arabic numerals; quaternary, by lowercase letters; and if needed, further minutiae indicated by lowercase Roman numerals. (Actually, I have seen some extensive old scientific scripts that used Greek letters.)

The urgency in pragmatic technology demanded by World War II had neither time nor patience for old classic indices. The simple, clear directness and convenience of Arabic digits made sequential numeric indexing the system of universal choice over other inconsistent modes. Really, "8.5.433" is easier to read and use than, "VIII, E, 4, c, iii," especially if you seek a cross-index item while doing a titration.

2.4.1 ADVANTAGES OF THE DECIMAL-NUMERIC WAY

In Figure 2.4.1, the right side is the outline for a proposal at a Brooklyn public service agency, properly indexed by the numbers. The left side shows what the same outline would look like in the Harvard (or any other mixed character) system. For instance, to find a reference, "2," would that be the "2" in II A? Or in II B? Might "a" be the "a" in II C 2? Or in II B 3?

In a corporate setting, using the decimal-numeric way, you can parcel out four separate parts of a large document to be written simultaneously by four different groups and keep track of the pages until final assembly for pagination. And should some mishap scatter the stack of unnumbered papers, they can be recompiled easily by following the decimal index numbers, which would be painstakingly difficult in any other system.

Once you have developed a firm numbered outline, you do not have to commence writing at 1.0. This method allows you to start writing anywhere

FIGURE 2.4.1. COMPARISON OF INDEXING SYSTEMS: HARVARD AND DECIMAL-NUMERIC

HARVARD (Archaic)	DECIMAL-NUMERIC (Current)
I. PREAMBLE	1.0 PREAMBLE
A.	1.1 Motivation
1.	1.1.1 Sponsors
II. COMMUNITY	2.0 COMMUNITY LIAISON
A.	2.1 Program Objectives
1.	2.1.1 Immediate
2.	2.1.2 Long Range
B.	2.2 Program Organization
1.	2.2.1 Steering Committee
2.	2.2.2 Community Boards
3.	2.2.3 Steering Committee Staff
a.	2.2.31 Executive Director
b.	2.2.32 Community Liaison & Public Information Company
c.	2.2.33 Local Area Planning & Design Teams
C.	2.3 Program Functions
1.	2.3.1 Community Liaison
2.	2.3.2 Public Information
a.	2.3.21 Press
b.	2.3.22 Radio & TV
III. SUMMARY	3.0 SUMMARY
A.	3.1 For Public
B.	3.2 For Sponsor

you feel comfortable (sec. 3.2) and file the finished parcels in sequence. Because the decimal-numeric system provides every entry a specific place, it is computer friendly, which is why most nonfiction publishers prefer it.

2.4.2

It is easy to go astray in other systems. Lack of early enlightenment, unwillingness to change, or plain laziness can lead to confusion in other systems, creating misunderstanding or misinterpretation. Figure 2.4.2 is a good example of the many ways to make a bad impression in any reader's mind. The sample is right from the engineering dean's office at a prominent university. It is part of a memorandum about an award. The outline is fraught

ORGANIZATION AND DEVELOPMENT

FIGURE 2.4.2.   HOW THINGS CAN GO WRONG

> 4.0   ~~IV.~~ *Award*
>   4.1   ~~a.~~ *suitable inscribed certificate or plaque*
>   4.2   ~~b.~~ *cash prize*
>   4.3   ~~c.~~ *subscription to* Scientific American
>   4.4   ~~d.~~ *expense paid trip for one-day seminar*
>
> 5.0   ~~V.~~ *Society responsibility*
>   5.1   ~~1.~~ *local school contact and selection*
>   5.2   ~~2.~~ *local media contact, with assistance*
>   5.3   ~~3.~~ *arrangement and presentation of award*
>   5.4   ~~4.~~ *coat of local activities*
>   5.5   ~~5.~~ *help fund awards, if more than certificates*
>
> 6.0   ~~VI.~~ *College responsibility*
>   6.1   ~~A.~~ *provide school and school personnel*
>   6.2   ~~B.~~ *prepare selection and procedural guidelines*
>   6.3   ~~C.~~ *prepare certificates and/or other awards*
>   6.4   ~~D.~~ *help initiate news release*
>   6.5   ~~E.~~ *help fund awards if more than certificates*

with errors of concept, indexing, and indentation. The columnar position of a notation has equal value with the notation.

Figure 2.4.2 shows the memo as it was issued, in standard type font; my edits are in *italics*. The difference is clear. If the original indexing symbols were correct, their columnar left marginal position would be wrong. In 4.0 (IV), the four entries, marked in lowercase letters, would belong in the fourth column, after material from an "A" and a "1." In 5.0 (V), those five items marked by Arabic numerals should indicate a third rule order and belong in the third indent after material in "A." In 6.0 (VI), the capital letter postings are incorrectly indicated, but in the proper columnar order. None of that confusion would have occurred if the postings I show in italics had been used. Things like that never happen when using the decimal-numeric way.

2.5 OUTLINE BUILDING

Another look at Figure 2.0 shows the separation between any two stages from P to W as a definitive line indicating a definite change of activity. That is *not* so in the one case between D and O. Not a printing smudge, that gray area conveys the notion of a gradual activity transformation: What begins as Data Distillation (D) metamorphoses to Outline Building (O), a smooth transition to enable one continuous process. This is your last chance to reshuffle any entity (key thought) to another place where it might be more effective in response to A. Do all this with great care; here is where your text body emerges.

2.5.1 THE SUBORDINATED SORTS

When you started this project (sec. 2.1.421) you accumulated an open list of questions about this subject toward being responsive to A. To break out from the confines of your own limited knowledge, you did *not* begin by listing possible categories to fill in. Instead you had the chance to ask unlimited questions to shake out facts you didn't know you didn't know. After collecting a reasonable number of questions (sec. 2.1.422), you sorted them into groups of like nature, then watched the category group titles rise from the commonality of those questions. That enabled you to find Plant, a group previously unrealized.

Now that you are past seeking additional possibilities, the exact opposite occurs in dealing with the *answer* face sheets. Here you have a finite number of facts in known subgroups, thus you may add *inert umbrella headings* as in-line thought name tags. Those umbrella headings do not add meat to the matter; being *inert*, they aid in achieving mellifluous movement.

2.5.2 OUTLINE DETAILING

The entire section 2.4 justifies the decimal-numeric system as the logical, sensible, practical method for organizing any writing effort. Section 2.5.1 explains the value of adding *inert umbrella headings* at this point in the project, identifying and packaging listed factual entries. Realize that what you do with any one face sheet, you must do equally, one group at a time, to all other face sheets. Closest at hand, you start with Marketing (fig. 2.3.3); that sheet has seven listed items.

5.0 *Marketing*
    eggs locally
    eggs trade
    fertilizer
    bedding feathers
    market area eggs
    chickens
    decoration feathers

You must rearrange this list for it to tell of its economic significance; add *inert umbrella heads*. To do that sensibly you choose to sort them according to things you can sell. Your prime commodity is *eggs*, the basic reason for this assignment. Next is *chickens*. According to the principal poultry pundits at Cornell University Animal Science, an average white leghorn hen's natural life span is eight to ten years, but menopause occurs at about age four, rendering the hen no longer productive. Since they don't make good pets, you sell them to Kentucky Fried Chicken. You lump the small percentage of other items into *byproducts*. Thus, rearranging the seven entries from the face sheet list in section 2.5.2 and *heading* them, your outline develops into a useable:

> 5.0 *Marketing*
> 5.1 *Eggs*
>   5.1.1 Local sales
>   5.1.2 Trade area
> 5.2 *Chickens*
>   5.2.1 Meat
> 5.3 *Byproducts*
>   5.3.1 *Feathers*
>   5.3.11 Bedding
>   5.3.12 Decoration
>   5.3.2 *Litter*
>   5.3.21 Fertilizer

2.5.3 SINGLE-ENTRY NOTATION

In that completed outline of 5.2 Chickens, there is only one subentry, 5.2.1 Meat. Some traditional teachers of English still cling to the notion that "If you don't have a 'b' then you may not have a lone 'a.'" No! There is nothing wrong with having a single subentry if that is where the item fits into the flow of events. Raising it a notch to conform to that unfounded rule erroneously puts it into a higher classification.

2.6 RECAPITULATION, OUTLINE RECASTING, AND VARIATIONS

You have just experienced the troika of applying the theory of chapter 1, developing a hypothetical assignment, and organizing your own report, all in parallel time. Here now is the evolution of a genuine project that demonstrates the flexibility this system affords for accomplishing differently targeted documents using the same acquired source material.

This next example is sanitized to preserve the propriety and privacy of all involved. It is nonetheless genuine. A local public utility wanted to investigate a recommended system's potential for improving the house operation. J. F. O'Hagan was assigned the task.

2.6.1

Having had this instruction, he knew the route well. O'Hagan noted his specific prime reason for writing: *To provide material to decide to use DART*; generically, *Information*. His second reason: *To document the task was completed*; generically, *Record*.

2.6.2

O'Hagan clearly defined his readers: (1) the *apparent* reader, the vice president in charge, the source of the assignment, and the addressee of record, is *archivistically inclined:* he will read, but not evaluate, the report; (2) the *real* audience, the operational supervisor, is an *action-oriented* person who will read and implement the report; (3) the auditor is a *secondary* reader who is interested only in the *fiscal aspects*.

2.6.3

Determining logically from his reasons and reader lists, O'Hagan titled his report "Design and Assignment by Remote Terminal, DART."

2.6.4

To decide what about DART all audience levels would need and want to know, he put himself into all readers' shoes, then generated a list of 24 random questions they might ask. Then he shook those 24 questions into groups of similar or like nature. That sorting resulted in four groups of kindred queries. Their likeness determined their group names as Advantages, Document, Makeup, and Immediate Considerations.

2.6.5

As an aspiration and a target marker for direction, O'Hagan wrote himself an SS (fig. 2.6.5) that he later used with slight modifications as the first paragraph of his final report text.

ORGANIZATION AND DEVELOPMENT

FIGURE 2.6.5. SCOPE SUMMARY PARAGRAPH

> **SCOPE SUMMARY**
> Can the Facility Assignment Bureau's Operation be improved physically and economically by employing a system of Design and Assignment by Remote Terminal? Data on all aspects of the suggested system makeup, advantages, effect on the present system will be collected, analyzed, and studied to enable an informed decision.

2.6.6

Planning, then going, he read, observed, and interviewed to get his questions answered fully. As he went, he sorted answer material into four labeled 10" × 12" envelopes and noted each entry on its envelope front.

He returned to his office, photocopied the envelope fronts to yield workable *inventory face sheets* (fig. 2.6.6), then stored the full envelopes safely, out of the way.

FIGURE 2.6.6. INVENTORY FACT SHEETS

Advantages	Document	
Real costs	Proofreading	
Paper costs	Production	
Time	Reproduction	Binding
	Local distribution	Reading
	Company distribution	Graphics
	In-house distribution	Text

Makeup	Immediate Concerns
F.A.B.O. Plant	Control functions in S.O.P.
F.A.B.O. General description	B.S.P. standard budget
F.A.B.O. Equipment	B.S.P. standard schedule
Definition	Impact on present system
Implementation	Pics control
	Pics record

2.6.7 PRIMARY SORT POSSIBILITIES

To build his outline from the material on hand, O'Hagan used the four *inventory face sheets* (fig. 2.6.6), his SS (fig. 2.6.5), and his reader priority note (sec. 2.6.2) to set a suitable firm order.

If he were targeting the *secondary* reader (finance interest only):

> 1.0 Advantages
> 2.0 Immediate Concerns
> 3.0 Document
> 4.0 Makeup. *Not Applicable*

If he were addressing the *apparent* audience (archivistic interest):

> 1.0 Document
> 2.0 Makeup
> 3.0 Immediate Concerns
> 4.0 Advantages. *Not Applicable*

If he were addressing his action-oriented *real* reader:

> 1.0 Makeup
> 2.0 Advantages
> 3.0 Immediate Concerns
> 4.0 Document. *The choice*

All of section 1.5 goes into detail about why the reader is (after "why write?") the second basic consideration of every document project. This section has just illustrated some possible variations for arranging the available material at hand to capture and hold a target reader.

2.6.8 SECONDARY AND LESSER OUTLINE SORTS

Once his primary order was established, O'Hagan sorted the entries within each, one sheet at a time.

2.6.81

Starting with 1.0, *Makeup* (fig. 2.6.6) has five entries. Logically, *Definition* seemed best to start, and definitely *Implementation* would close it with "teeth." The three remaining notations all deal with FABO, but *no* need to make an inert umbrella heading because one of those three was the *General Description*,

FABO. Subordinated to that was the *Plant*, housing the operation. The *Equipment* within it was further subordinated. He used all five entries in this outline:

>   1.0 Makeup
>   1.1 Definition
>   1.2 General Description, FABO
>   1.2.1 Plant
>   1.2.11 Equipment

2.6.82

His second face sheet, *Advantages* (fig. 2.6.6), had three entries, of which *Time* was the main idea. The other two were about *Costs*; thus, he inserted a combining inert umbrella heading under which he listed *Paper* and *Real* as equal weight notations. This time three notations yielded four entries:

>   2.0 Advantages
>   2.1 Time
>   2.2 Costs
>   2.2.1 Paper
>   2.2.2 Real

2.6.83

Using the *same* logical procedure to organize the two remaining face sheets and combining all four sections, O'Hagan had his firm outline:

>   3.0 Immediate Considerations
>   3.1 Impact on Present Systems
>   3.1.1 Plug-In Coordinator
>   3.1.11 Control
>   3.1.12 Record
>   3.1.2 Control Functions in Practice
>   3.2 Standardizing via Utility System Practices
>   3.2.1 Budget
>   3.2.2 Schedule
>   4.0 Document
>   4.1 Document Description
>   4.1.1 Reading
>   4.1.2 Text
>   4.1.3 Graphics

        4.2 Production
           4.2.1 Reproduction
               4.2.11 Proofreading
           4.2.2 Binding
        4.3 Distribution
           4.3.1 In-house
           4.3.2 Local

From that he wrote his successful report, which was responsive to the charge in A and met the aspiration of his SS.

To see another subject's fully formatted complete report, see appendix 9.2.1.

2.6.9 DETERMINING ORDERS OF SUBORDINATION

In section 2.6.81, the order of the outline was set because the principal topic was *Makeup*. In that, *FABO* occurred in a reinforced concrete *Plant* building, where a piece of mobile *Equipment* functioned. Their outline positions were determined by their interrelations on the ground. That was the case for this report, but not always.

2.6.91

For instance, consider an oil refinery as your subject. In that, there is a catalytic "cracker," a huge, stand-alone, outside piece of *Equipment*. Hung onto it, a series of catwalks, inspection ladders, and lighting fixtures are its *Plant*. That outline (last two items now reversed) would appear as

        1.2 Catalytic "cracker"
           1.2.1 Equipment
               1.2.11 Plant

2.6.92

Another variation based on interrelationship involves an automotive assembly facility. The plant building was designed to house the machinery; and that was designed simultaneously to fit the building. That report outline would be

        1.2 Assembly Facility
           1.2.1 Plant
           1.2.2 Equipment

Again, order is determined by the facts in situ and facilitated by the logical flexibility of this system.

2.7

In this long chapter 2, I've laid out step by step the applied theory, a hypothetical project, and two real project developments, each evolved from the Assignment (A) to Dispersion Point d, the start of the Writing stage (W). Now you should have at hand for your own document your *reason(s)* for writing, to set the tone; *targeted audience(s)*, to set the levels; *Scope Summary*, to define your goal; *researched material* (in envelopes or on your computer) to provide the substance; and a *firm outline*, your actual report without the details. With those five items you have all you need to write your report. And you will write confidently after having worked through chapters 3, 4, and 9. When your draft is finished, chapters 5, 6, 7, and 8 will provide you with the "cheese on the spaghetti" to make your document polished and professional. Complete two or three more projects to become *comfortable* with this system. Take your time with it: "Haste, haste has no blessing."[2] Become familiar with the finer facets. There are no prizes for getting there first. The trick is getting there *with ease*.

Then start with chapter 10 and work your way on to the book's end. Confidently, with those acquired, honed, and polished communication skills, you should be worth thousands of dollars more per year to your client or employer, and to yourself.

WHERE THE NEEDLE GOES, THE THREAD WILL FOLLOW.[3]

NOTES

1. Once, one young man in Baltimore responded, "I'd throw it on the floor."
2. Swahili proverb.
3. Old Russian proverb.

3
WRITING TECHNIQUE

3.0

This is the way to avoid staring at a blank screen, or a blank page. You *can* get past the paralysis that parries pen from paper. After you have collected all of everything (Scope Summary, completed current outline, and researched reference material) required in section 2.5, take a short breather, then whack in and start to write. The writing is the actual end product of all your preliminary planning, research, and construction. It should be as nearly perfect as your professional pride demands.

3.1 OPTIMUM ENVIRONMENT

To write best you need comfortable surroundings, an ambient atmosphere, and the fewest external distractions. Seldom does a writer enjoy the delight of absolutely perfect writing conditions. But you can manage to approach them via many of these suggestions.

3.1.1 WORKPLACE

Best is in a private office or room. Next is a cubicle. Most difficult is in an open office where you are vulnerable to everything going on around you. Try for isolation and emotional privacy; at least no interruptions.

3.1.2 QUIET

Outside noise, pop music, and the news infiltrate your concentration. If all else fails, try earplugs or cotton. Depending upon your disposition, modulated classical music may be relaxing. Diminish it if it makes you listen instead of just hearing it.

3.1.3 CHAIR

Your chair should be posture supporting and comfortable, but not soporific.

3.1.4 LIGHTING

Lighting should be neither bright nor dim; either strains your eyes, may cause headaches, and will shorten your writing time.

3.1.5 IMPLEMENTS

Check if your computer is clear and in prime working order, or make sure you have a supply of sharpened pencils and full pens.

3.1.6 REFERENCE MATERIAL

All your reference material should be close by and readily accessible without your having to move far.

3.1.7 REFRESHMENT

Having a water bottle, or cup of tea or coffee handy for an occasional sip helps. Going after it is a time waster. Be careful to put the container where it will not be tipped and spilled on your papers.

3.1.8 TIMING IS PARAMOUNT

Good writing takes careful, in-depth time. Allow several hours to write a major effort, or at least two hours to do a lesser item. It should be solid time, not several spurts of a few minutes in your calendar interstices. A book or long report calls for protracted time. Write it by scheduling the same time slot every day, and staying with it. That time allotment must be uninterrupted to allow for your total concentration.

3.2 THE PLUNGE

This is the time of truth where you mix all the ingredients (sec. 3.1) into a solid product. You do not have to start at "1.0 In the beginning . . . " If you have compiled your outline by the numbers (sec. 2.4), and have your Scope Summary (sec. 2.1.43) as the target marker—your objective—you may and you can start anywhere and file it by the numbers. You begin where you wish, but you must write the entire section. Never a bit here and a bit there. Paragraph jumping serves to create your own disorganization.

Among your outlined sections there must be one that you know best, or like best, or with which you are most familiar. That is a good place to start because you can write that the easiest, and it will yield you one drafted chapter.

Now look at your outline section headings and set yourself an order of precedence. You may make it the next easiest, and the next, until all your easy or familiar sections are drafted, then tackle the more difficult ones. Or you may draft the toughest ones first; that done, write the easier ones going downhill.

Any of those routes should deal with all the interior sections in any order from 2.0 to your penultimate one. But do not write either the opening section, *1.0*, or your last section. When you have drafted all the interior sections, then write your summary closing section to keep your material flow from "falling out the bottom." After all other sections are drafted, *then* you may compose your opening section, *1.0*. By being familiar with the whole work, you may write a realistic introduction to a document in being. If you do not write it last, you may promise something in your opening section you might later omit or neglect in subsequent draft development.

3.3 WRITING

Whether you tap it out on your laptop, or write it with pen-in-hand on paper, the process is the same, and you waste no time wondering where to start.

3.3.1 FIRST DRAFT

Just as if you were telling it to your receiver over a cup of coffee, in plain words, write everything out. Get as much wordage as you can onto paper while it is coming to you. No interruptions, no stopping to check grammar, or spelling, or punctuation. Even if you put it all down in one continuous sentence, get it down without a timeout. It should flow easily because your outline (sec. 2.4) has been well constructed, and you head for the expected goal you set in your S. S. (sec. 2.1.4).

In the past, in the age of film and hand-adjusted cameras, I went bird watching with an avid photographer. Silently, we spotted a rare Blackburnian Warbler, perched in plain sight. While the man fiddled with his light meter, the nervous bird took off. Had he shot on sight on universal focus, later he could have doctored his film in the darkroom. But he had nothing. Commit all your guided thoughts to paper first; later you can adjust, shift, and refine a draft in being.

3.3.2 SECOND DRAFT

Now you have an overstuffed rough, rough draft unchecked for solecisms. But it is organized and contains all your necessary information. Start by shaping it into sentences while casting out redundancies, eliminating unnecessary words ("what is it that he wants" becomes "what does he want"). Tighten the sentences then spell check the whole reduction carefully, and punctuate it properly.

3.3.3 THIRD DRAFT

The third draft is a clean rewrite of the second, without any encumbrances or solecisms. If you work in an organization, this is the one you submit for superior approval, if demanded by protocol. If you are an independent, you might wish to try this on your editor, or a colleague.

Select graphics (sec. 5.3) to enhance the written text. For some magazine articles, colorful graphics make the piece more attractive, to entice readers. Be sure you caption and reference every graphic to its related place in the text.

3.3.4 FOURTH DRAFT

This draft is only required if you need one more polishing of the third draft, just to be sure. If you were reading it, not writing it, is it what you'd want to see?

3.4 TAPE RECORDER ALTERNATIVE

Some writers talk better than they write. Only experience will tell you which you are. You must be comfortable when setting out the first, free draft. If you choose to dictate, then tell it to the recorder as it is, as it comes to mind, then transcribe the recorded text onto a word processor. Deal with your double-spaced printout as a first draft following the route of sections 3.3.1, 3.3.2, 3.3.3, and 3.3.4. Either section 3.3 or 3.4 should yield you a good piece of work; go do it.

THERE IS NO MAGIC MEDICINE OF ONE DOSE.

4

CHARACTERISTICS, REFINEMENTS, AND STYLE

4.0

You should weave these characteristics and refinements into your style as you write your document. Regard your reason for writing (sec. 2.1.2) and your audience (sec. 2.1.3) as the determining factors of what you put into your text. Write it all in first- to second-person active voice. Your document can be more formal (a report or a journal piece), or more open (an item for a magazine or newspaper). Here are the characteristics you must impart to either type of document you turn out:

4.1 MORE FORMAL TEXT (REPORT)

4.1.1

Formal text is addressed to a real, named audience (secs. 1.3.2 and 2.1.3).

4.1.2

Be responsive to the charge. You must prove (or disprove, or detail) a specific question to a specific end.

4.1.3

Use graphics fully (sec. 5.6) to augment (not replace) the written word.

4.1.4

Include your personal opinions only if requested by the assignment.

4.1.5

You must cite and define causes and effects (sec. 8.2.3).

4.1.6

The text should be tabulated using the numeric-decimal system (the Harvard System is not a good option) (secs. 2.4 and 2.4.1).

4.1.7

Write in crisp, declarative, brief sentences, observing all rules of grammar, punctuation, spelling, and sentence structure (sec. 4.3.1).

4.1.8

The conclusion should answer the original charge of the assignment.

4.1.9

Find suggested formats in chapter 9.

4.2 MORE OPEN TEXT (PUBLICATIONS)

4.2.1

The text is, in effect, addressed "to whom it may concern."

4.2.2

The text need only involve a central theme for general interest, but it can be all-inclusive.

4.2.3

The text is usually narrative; graphics are mainly for dressing.

4.2.4

At your option, the text may or may not include your personal opinions.

4.2.5

You must cite and define causes and effects (sec. 8.2.3).

4.2.6

The text is narrative, but it may include interspersed headings.

4.2.7

Use your best narrative style.

4.2.8

A formal conclusion is at your discretion; you may not want one.

4.2.9

Suggested formats are in chapter 15.

In the previous comparison, item 4.1.6 advocates the numeric-decimal system for indexing documents, as it is superior to the old-fashioned Harvard System. The latter uses two types of numbers and two cases of letters. That recommendation and its implications were discussed in detail in section 2.4.1.

4.3 SOME HELPFUL SUGGESTIONS

Here are some things I've learned over the past half century that you may not find in other books:

4.3.1 NO SOLECISMS

Follow all the rules of English grammar, sentence structure, and punctuation. You might have paid more attention in English 101, but now you can get good help from Strunk and White, *The Elements of Style*, any edition, then enjoy the satirical list in appendix 4.3.1.

4.3.2 SPELLING

Spelling is important. I cannot spell every English word, but when reading, I can tell bad spelling. A misspelled word jumps off the page and punches me in the eye. Bad spelling can deflate any professional letter because it is readily detectable and gives a negative impression of you in the receiver's mind. Reliance on a computer's spell checker is not the best bet. Sometimes it is wrong. Among others over the past few years, an instance quickly comes to mind. The spell checker responded to my inquiry for *supersede* with *supercede*. Wrong. It is from the Latin *seder*, to sit (above).

Don't overlook the obvious; some misspellings may be hiding in plain sight. And some correct ones may not appear at first sight. Check every doubtful word. "Wright, write rite right," is a correctly spelled and punctuated sentence that could have been written as, "Wheelmaker, escribe protocol correctly."

The coach pushed the dean to restore eligibility to his suspended defensive tackle, for Saturday's big football game. Reluctantly the dean offered the player, "This is a spelling test Joe, I'll give you one word, and if you get just *one* letter correct, you pass. Spell coffee!" "Uhh, K-A-U-P-H-Y."

You all know about phonetic variations. One spelled fish as GHOTI because he learned *f* as in touGH, *I* as in wOmen, and *sh* as in noTIon.

Don't forget the *e* between the *g* and the *a* in *interchangeable*. And no *e* between the *g* and the *m* in both *acknowledgment* and *judgment*. Don't forget one *s* and two *c*'s in desiccate. Make yourself use the dictionary. Not only will you get it right, now, but you'll remember it for next time.

4.3.3 PUNCTUATION

All English grammar texts will give you the rules for standard punctuation, but they do not give enough attention to the use (or misuse) of the hyphen (-). It is used correctly when breaking a word at the right margin that will continue on the next line, and it is proper in combining words, as in *old-fashioned*. Lazy writers often use a hyphen in place of a comma. Wrong! The hyphen and the comma are *not* interchangeable. Actually, they tell the reader diametrically opposite things. A comma is vertical, like a partition, *separating*. A hyphen is horizontal, like a coupling *joining* two railroad cars. The word hyphen is from the Greek *hyph'hen*, it means together. Here, see the opposite meanings (app. 4.3.3):

Try caffeine-free Coke.
Try caffeine, free Coke.

A comma in the wrong place can totally *reverse* the intended meaning of a sentence.

Surrender, impossible to fight on.
Surrender impossible, to fight on.

In a series, items should be separated by commas, including one between the penultimate item and the *and*.
We had melon, cereal, toast, and coffee. The last two are separate items; *do* use a comma.

We had melon, toast, coffee, and ham and eggs. The last two are paired items; do *not* use a comma.

And try this one, and see the difference: *Men, Women, and Dogs.* "*Men, Women and Dogs*"[1] (app. 4.3.3).

4.3.4 SENTENCE LENGTH

Try to keep your sentences to about a dozen words, *average*, but not rigidly so. Don't add to your work problems by counting words; it wastes time. If you need a fifty-two-word sentence to explain superheating, then use fifty-two. And there is nothing wrong with a one-word sentence such as "No." Use your judgment.

4.3.5 REDUNDANCIES

Omit using nonworking, repetitive, redundant words over and over again. They take space and waste the writer's and reader's time. They add to production costs and bore readers.

Not all deadheads are repetitions. There are many words that do not add to the meaning of a statement, but do bloat. A few "don'ts" in common use that need avoiding:

In other words. If you don't think a phrase you've written states your intended thought, do not let it stand, then add "in other words," followed by a restatement that may or may not be that original bit *in* other words. Do cancel your weak phrase, don't use "in other words," then state your better version. Why waste space confusing the reader? In other words, don't use it!

In order, or *so as*. Before every infinitive "to . . . " the *in order* is useless fat. Without it the statement is clearer and unencumbered. *In order to read, I need my glasses*: declare the "in order" out of order; omit it. *To read, I need my glasses* says the same thing in fewer words. This same argument goes for *so as*.

Located. By eliminating the useless *located* in all your writing, you save time, space, and materials. If you dislocate the located from *Boulder Dam is located in Nevada*, you have the shorter and cleaner *Boulder Dam is in Nevada. The office is ~~located~~ in room 313*. To locate is a double-edged word that can mean either *to place* or *to find*; opposite intentions. *Locate the gauge located in the kit* can mean either *Find the gauge placed in the kit*, or *Place the gauge you find in the kit*. Do not leave your reader wondering. Depending upon the direction of the sentence, use either alternate word to be clear. Yet another complication: Recently an acquaintance phoned and said, "Morris, I want to locate a young woman." I responded, "Have you tried Missing Persons?" "No you clown, I want to get her a job." You do not need *located* in conjunction with *any* form of the verb *to be*.

The following or *as follows*. If a list or explanation is the next thing after a colon (:), then you do not need either of those expressions before the colon. A colon is a flashing red light that proclaims something follows immediately hereafter. It speaks for itself, eliminating the need for *the following*, or *as follows*. *To start writing a novel, you need the following:* can be written more succinctly as *To start writing a novel, you need: . . .* Further, *you can cut the deadheads as follows:* by just writing *To cut the deadheads:*

Period of time. A quart can be the measure of bourbon, or ice cream, or engine oil, and thus must be so labeled. But period is a unit that measures only time. In common use today is the too often *period of time*. You need only one, either *I've been teaching for a long period* or *I've been teaching for a long time*. Either, or, but not both. And *at this point in time* is best written as *now*.

Another gem of the same ilk: *cancel out*. If you cancel it, it's out!

At the start of a book, some publishers head the list of chapters as *Table of Contents*. Obviously it is a table, ergo just *Contents* is enough.

Raise it up or *lower it down*. Ever see anything raised or lowered sideways?

And, *needless to say*, is needless to say.

4.3.6 MANAGEMENT OF WORDS

Your management of words in a sentence has as much to say as the words themselves in the clear, correct conveyance of an idea. Careless or lazy transposition in the old chestnut "I saw the fire running around block" really means "Running around the block, I saw the fire." But the transposed original statement could be arrested by adding *while* to read its intended meaning: "*While* running around the block, I saw the fire."

A while ago, at an editorial meeting, the woman's work editor proudly displayed the jacket of her latest release, *The Effective Woman Manager*. I asked her if that were a book about a chap who managed his women with ease. She went up in a puff of blue smoke because in her haste to get the manuscript into print she neglected to see into her good title the possible malthinking of chauvinists. By rearranging those four words to *Woman, The Effective Manager*, not only would it have eliminated the erroneous misdirection, but it set *Woman*, the subjective word, as the title's first word.

Bring and take are not interchangeable. Properly each indicates the direction the subject moves in relation to the destination. Bring *comes in*: *bring home the bacon*, or, *bring in the milk*. Take *goes out*: *take this to the bank*, or *barbecue take out*, or *take me home*.

Less and fewer. Less deals with groups or collective nouns; fewer deals with individual units. *Less traffic*, but *fewer cars*; *less money*, but *fewer dollars*; *less time*, but *fewer hours*, or "*Europe is the less . . .*"[2]

Those refers to the past. Use it after material that has been done or finished. *Those* were the days . . . *All those* were accomplished by nightfall.

Never utilize *utilize* when you can use *use*; and *lengthy* is a lengthy way of writing *long*.

These, now or future, is used in place of the eschewed "the following" or "as follows" when preceding a statement about something present or future. *These* are the tools available. *These* men will become eligible. . . . If you write about something past or earlier on the page, use *those* in place of *these*.

In your final manuscript copy, be absolutely sure the *position* of a word will make the sentence say what you mean. Take a simple, clear declaration:

I drank my daughter's beer.

Five words, six possible places to add an extra word.

(1) *I* (2) *drank* (3) *my* (4) *daughter's* (5) *beer* (6).

Now add the single word *only*, once in each space.

(1) *Only* I drank my daughter's beer.

That means *no one else did*.

(2) I *only* drank my daughter's beer.

That means *that's all I did*.

(3) I drank *only* my daughter's beer.

That means I drank *just Misty's, no one else's*.

(4) I drank my *only* daughter's beer.

That means *Misty has no sisters*.

(5) I drank my daughter's *only* beer.

That means *Misty had no other beers*.

(6) I drank my daughter's beer *only*.

That means I drank *just the beer, no kippers, no popcorn*.

4.4 AVOID COMPLEX OR COMPLICATED SENTENCES

Be simple but not simplistic. (One is an easy read, the second is patting the reader on the head with juvenile discourse.) William of Occam (Ockham), "The Invincible Doctor" of fourteenth-century England, sagely advised, "Be simple in your explanations, don't multiply entities beyond necessity."[3] That sagacity is as true today as it has been over the past eight centuries. And Leonardo da Vinci, probably the most brilliant "engineer" of the Renaissance, wisely opined, "Simplicity is the ultimate sophistication" (app. 4.4).

4.5 AVOID JARGON AND COLLOQUIALISMS

Avoid jargon and colloquialisms, "institution-itis," hedging, and vague, noncommittal, or imprecise statements; be positive (see app. 4.5.1 and 4.5.2).

Some years ago Lieutenant Barton S. Finegan, Department of Naval Science at Cornell University, asked me to edit his annual report. Read both versions (fig. 4.5) and see how the reader-friendly version imparts the same facts in a shorter, far more accessible way.

Often have I advised: know your receiver, then write to suit both his reception levels (i.e., technical and comprehension). But you must realize that he might not be the only interested party. Do not "dumb down" for everyone's understanding because then your message may become so diluted that no one will understand. If you write on a specific aspect of a highly technical topic to a pinpointed reader, it is all right to be as complex as the conditions demand. For example, on page 14 of *The Energetics of Mitochondria*, J. E. Chappel (Ed. by J. J. Lead) writes,

> Thus, NAD+ is reduced by dehydrogenases in the tricarboxylate cycle in the matrix of the mitochondrion, and the resulting NADH is oxidized by the NADH dehydrogenase on the christae.

That paragraph is in its closest compass, no deadheads, no superfluous words. It is succinctly clear to its target audience. No one except they will read, understand completely, and be able to use that information properly. It will not matter to anyone else.

In contradistinction to the previous Chappell example, a message addressed to the general public must not be written in institution-itis. Because the audience is unlimited, both receiver levels are unlimited also. Do not assume everyone is "hip to your jive," nor even "in sync" with it.

FIGURE 4.5.

Archaic (Impersonal-Passive)	Current (Humanistic-Active)
The NROTC Sail Training Program at this University has matured into a most professional and successful endeavor. Voluntary integration with the Athletic Department's Sailing Program has fostered a beneficial interchange of information and resources. Six damaged and aging sailboats were borrowed from the Athletic Department, of which five have been refurbished and are presently operational on Cayuga Lake. The training provided on those boats has become a welcome addition to the "Big Boat" training which is conducted aboard China Doll, the NROTC Yacht. Outfitted with the latest equipment, and capable of embarking a crew of 12 for extended periods of time, China Doll has proved to be an excellent vehicle on which many of the following skills such as navigation, seamanship, shiphandling, and leadership can be taught. China Doll is at the present time cruising the Atlantic coast with a crew of officers and midshipmen visiting various ports of call. It should be noted that this type of training for our future Naval Officers is certainly a superb complement to the NROTC academic program. (175 words)	The NROTC Sail Training Program at Cornell has matured into a most professional and successful effort. Our students have worked closely with the Athletic Department's Sailing Instructor. This achieved beneficial exchanges of information and resources. We borrowed six damaged and aging sailboats from the Athletic Department. NROTC students completely refurbished five boats, and are now sailing them on Cayuga Lake. The training we provide on those boats is a welcome addition to the "Big Boat" sailing aboard the NROTC Yacht, China Doll. Outfitted with the latest equipment, and capable of embarking 12 for extended periods, China Doll is an excellent vehicle aboard which students can acquire skills in navigation, seamanship, shiphandling, and leadership in a real environment. China Doll is now cruising the Atlantic coast, with a crew of officers and midshipmen, visiting various ports of call. We believe this sort of training for our future Naval Officers is a superb complement to our academic program. (153 words)

Brooke Knapp is the first pilot to have flown around the world via a great-circle route that naturally took her over both poles. En route she made an emergency stop at Recife, Brazil. She later described it to the world via the public press in pure jargon. The San Jose (California) *Mercury News*, for Wednesday, November 23, 1983, quoted her as saying,

> We took off out of Recife and selected "gear-up," and the mains came up but the nose wheel didn't, and we decided that wasn't a very good configuration for flying.

So we tried to recycle the gear. We put the landing gear selector "gear down." The gear did not go down so we had to blow it down with the alternate extension.

Interesting: In today's vernacular, "gear-up" could mean *prepare well for*; "good configuration" connotes *nice shape*; "recycle the gear" calls to mind *salvage it for materials*; "gear down" is *what you do to save wear on vehicle brakes*; and "blow it down" is *what high winds do to signs*.

There was no need for such insider lingo. That quote is too high-tech, aimed at postgraduate, seasoned specialists. It should have been accessible to everyone.

YOUR FINISHED DOCUMENT MUST REWARD YOUR EFFORT.

NOTES

1. James Thurber, *Men, Women and Dogs* (New York: Harcourt, Brace, 1943).
2. John Donne (1572–1631), *Meditation XVII*.
3. "*Entia non sunt multiplicanda prater neccessitatium*" (attributed); "*Frustra fit per plura quod potest fieri per pauciora*" (confirmed).

5
DESCRIPTIONS

5.0 DESCRIPTIONS

Descriptions, comparisons (chap. 6), and motion (chap. 7) constitute the three basic dimensions of all professional and technical communication. A good description of a description is a *word picture of an entity*. That entity could have mass (solid, liquid, or gas), or it could be an intangible (a condition, an idea, or a concept). It should not be drowned in words. Without nonverbal material or graphics, a properly worded description should create in the receiver's mind an identical image of the entity envisioned by the sender. Do not let the magic of today's sophisticated technology mislead you. There will be times when none of it is available and you must communicate an accurate description. You *can* do it solely with words. It is not the quantity of information you put into a description, it is the choice of nonambiguous, relevant, key facts you *do* include that make a useful description.

Everyone knows the old chestnut about the group of blind men, none of whom had ever encountered an elephant. One ran into a leg and thought the beast was like a tree. In turn, one grabbed the trunk, concluding it to be like a snake; another grabbed its tail, imagining it to resemble a bell rope; another was poked by the tusk and believed it to be like a sword scabbard; and so on.

Only when they all got together and shared views were they able to build a composite mental picture of an elephant. You have to view the whole subject from all angles, with perspicacity, so no aspect is overlooked, ignored, or overdone. And you all have to speak the same idiom.

5.1

A well-written external description of a process tube by Gordon Noyes with the Norton Company, Worcester, MA, is

> Process Tube, 5½", used as gas-flow kiln furniture by the semiconductor industry in the preparation of silicon discs for printed circuitry. The tube alone weighs 42 pounds. It consists of silicon carbide, and is slate gray with small, mica-like shiny spots covering the entire surface.
>
> The first segment has a 6" outside diameter, an inside diameter of 5½", and is 87" long. Fixed axially to one end, segment two is a 4" long, cone-shaped cap that tapers to an opening 1½" outside diameter by 1" inside diameter. Segment three, extending axially from this, is a 6" long tube, 1½" outside diameter by 1" inside diameter. The fourth segment is a 2½" outside diameter ball-joint that opens forward into a 1" inside diameter hole. The last three-piece assembly looks like an elongated nipple.

Can you get a clear mental picture of a process tube from Noyes's word picture? Try to sketch it yourself now, then turn to appendix 5.1 to see how close you came. Don't peek. J. T. Tocci of Letterkennny Army Depot read and envisioned the text, then produced the realistic sketch you see in appendix 5.1.

5.2

Although description writing fundamentals (sec. 5.7) must be observed, descriptions of tangibles may vary depending upon the writer's viewing position. But descriptions of the insides of a subject (a tank, a chamber, a space capsule) become easier because your point of view is locked into your surroundings—a Jonah-and-the-whale-type relationship. Ted Siegal of the Corps of Engineers, Baltimore, Maryland, was assigned to draw the interior and existing conditions of Room G-26A in the George Fallon Federal Office Building, with the objective of planning its reconfiguration. In an emergency, this written description was transmitted hastily to an architectural office where a draftsman easily transliterated Siegal's word picture into a working drawing. Try to sketch your interpretation of this memo, then check it against the office drawing that appears in appendix 5.2.

> Room G-26A, George Fallon Federal Bldg., Baltimore, MD. Rectangular, 30' E-W × 40' N-S × 9' high, with 4' N-S × 15' E-W offset projecting into southeast corner. No windows.

Reinforced concrete framing with columns at NE and SW corners projecting inward 6" N-S and 1" E-W. Columns are also on E and W walls 12' from S wall. W wall column is 2' wide, projecting 6" inward. E wall column is 4' wide and projects 1' inward. W wall is concrete block, plastered. N wall is full-height factory-finished metal panels. S wall and SE corner offsets are full-height factory-finished metal panels. S wall portion is acoustical rated vinyl accordion pleated partition. One 3' × 7' solid core hollow metal door, centered on the 15' face of the SE corner offset, opens to vestibule off outside corridor. A second door (same size and type) on S wall, 13' from the W wall, opens to storeroom. Both doors swing inward to G-26A. The corridor door is hinged on right when viewed from the room, and the storeroom door is hinged on the left.

Floor is reinforced concrete, carpeted wall-to-wall with synthetic fiber gold-tone carpeting over foam rubber underpadding.

Ceiling is suspended acoustic tile with recessed lighting fixtures. Hung framing is exposed "tees." Tiles are 4' N-S × 2' E-W. Light fixtures, 9" wide E-W × 3' long N-S, are 4' o.c. in alternate N-S rows of ceiling tiles, except along room center-line. Each fixture has three standard fluorescent tube units. Light switches on N wall up 4' 6", 3' from W wall and just E of corridor exit. Two electrical outlets on the W wall 10" above the floor, 4' and 20' from N wall.

Air conditioning vents surround each light fixture. 14" diameter return air inlets in ceiling on the N-S center line 13' and 26' from the N wall.

A 16' long × 5' high chalkboard is center mounted on the N wall; and another (same size) is center mounted at a point on the W wall 10' from the N wall. Bottom of chalkboard is 3' off floor. Pull-down projector screens are in 10' long housings centered over each chalkboard.

Note his word economy. "No windows," instead of perhaps, "There are no windows in the room," or variations, all overwritten. He is specific about door swing directions; tying left and right by stating, "When viewed from the room." Since he orients the room in the first paragraph, he could have written directions instead.

5.3 STYLE

Technical writing need not be dull, nor should it be so truncated it needs a guide. Until now I have advocated the precise, clipped style in description writing for short reports, memoranda, letters, and attachments. In writing for magazine articles, books, and news releases, present your facts in a reader-friendly but professional style. That is not a license to wax flowery, just open your wording carefully. Do not digress onto tangents. Keep your paper always moving toward your objective (see sec. 14.4).

In a long report about "Dredging in the Contaminated James River Channel," Richard L. Klein of the Corps of Engineers, Norfolk District, describes his subject and its surroundings in unmistakable, direct terms that are easily comprehensible and tightly stated. He is completely professional, yet he delivers his technical message as mellifluously as any good novelist.

> Dredging in the Contaminated James River Channel. The James River is a major tributary to the Chesapeake Bay. From headwaters in the Allegheny Mountains to rapids at Richmond, the James is a free-flowing stream crossing central Virginia. Below Richmond the river is tidal and navigable to its mouth. From Richmond to Hopewell, steep bluffs rise at the banks to the numerous industrial plants confining a narrow, winding channel. Below Hopewell the river grows wider, and the bends less severe. This lower leg of the James flows past colonial plantations, farms, historic settlements, and through the region's greatest seed oyster grounds. Finally it widens to several miles where it joins the deepwater ports at Hampton Roads.
>
> Congress authorized the James River Channel by passing the 1930 River and Harbor Act. We maintain minimum dimensions: 25 feet deep by 300 feet wide from the mouth to Hopewell; 25 feet by 200 feet to Richmond Harbor, the head of navigation.
>
> The channel serves the ports of Richmond and Hopewell and industrial uses such as DuPont and Allied Chemical. A variety of ocean-going and coastal vessels carry raw materials or finished goods to and from those facilities. Transportation savings and vessel safety are the primary benefits we gain from the channel.
>
> Sediments carried by the James River waters tend to settle in the channel at several places, forming shoals that obstruct navigation. Over the years, the Corps of Engineers has observed a pattern, thus were able to identify and name the shoals, then plan for their periodic removal by dredging.

As in many waterways, maintenance of the James River Channel was well suited to the cutter suction dredge. Above Hopewell, where the river is narrow, the dredged material was pumped economically into upland disposal sites. In the wide reaches below Hopewell, it was more economical to deposit the material into open water outside the channel. Engineers used their many years of dredging experience to choose the disposal sites that caused minimal channel resedimentation. At some places the disposed material built up, creating islands; at others it tended to disperse without reentering the channel. Today the conventional and most economical dredging method in the James River is still the cutter suction dredge.

5.4 RECEIVER LEVELS

Recall an earlier discussion (sec. 1.5.5) about writing to accommodate your reader's levels, *both* technical and comprehension. In a memorandum about a newly developed torque wrench, you would couch specific design details in different words to a shop mechanic from those for a professional engineer. Both men have significant roles in the wrench's production, but at different levels because of their training. Yet both must receive the same vital information. The answer is in your choice of words, and in sentence structure. Refer to basics in chapter 1.

To illustrate this idea, I asked Dr. Allyn B. Ley, a professor at Cornell University's School of Medicine in New York, how he describes shock to his first-year medical students. Then I asked Barbara L. Smith, RN, an industrial nurse at The Norton Co.'s Worcester, MA, plant how she would explain shock to a worker whose buddy had just come around from one. Both medical professionals defined the condition accurately, both accomplished a successful message, yet the wording of each was completely different.

> *To first-year medical students*: Shock is a clinical condition characterized by signs and symptoms that arise when the cardiac output is insufficient to fill the arterial tree with blood under sufficient pressure to provide organs and tissues with adequate blood flow. The low-flow state in vital organs is the common denominator in all forms of shock. It may arise from reduction in total body blood volume; that, caused by dilatation of the vascular bed from neurogenic causes, or from inadequacy of the pump function of the heart.

To manufacturing plant personnel: Shock is a systematic depression due to the lack of blood supply to the brain. It is a significant enough jolt to the body to change physical functions and appearance. Shock, caused by injury, can be identified quickly. The victim is weak and limp, the skin is pale, and the lips may be bluish. Breathing may be shallow and the pulse beats (easily felt at the side of the neck) are rapid and sometimes irregular. If loss of blood has occurred, you may observe restlessness, thirst, and anxiety. In the late stages of shock, the victim may be unresponsive, and both eye pupils may be widely dilated.

5.5

Describing *an idea, a notion, or a concept* is different from describing a physical entity. There is nothing you can see, hear, smell, touch, or taste. Ideas exist solely in the sender's mind, thus the job here is to get your receiver to comprehend your point of view, your thought, and its benefits. It is not easy to get a receiver to wrap his mind around a creation of your mind without a clear description. Too many bright people become mired in unorganized messages and fall to badgering a receiver with repetitions of "Did you get that?" or "See what I'm saying?" Not necessary.

Here is the easiest, most direct way. Think through your entire notion, and understand it all yourself. Then write a first rough draft. Rework that draft several times to boil out the nonessentials and deadheads. Be sure your basic premise is obvious, and that you have omitted the minutiae.

A local newspaper phoned Westinghouse Canada, Inc., Hamilton, Ontario, seeking a description of a contract item. T. Nanba was especially adroit in composing this descriptive reply, especially since English is his second language. He followed the suggested easy way.

> Third party maintenance services are those purchased by an organization other than the manufacturer/vendor, for the purchaser/end-user.
>
> Installation and maintenance services are important considerations in the successful marketing of telecommunication equipment, particularly in the field of airline reservation networks. Local sites can be serviced from our plant, however, cost and response factors limit our ability to serve distant, widely scattered American cities. There exist several independent firms who can, with training, provide such services as a "third party" to our clients. We may hire

them on a "demand" basis, wherein they will charge at a relatively high hourly rate, but only for services performed, or on a "contract" basis, wherein they will maintain equipment at a fixed monthly fee.

5.6 GRAPHICS

"One picture is worth more than ten thousand words," allegedly is a Chinese proverb according to Bartlett's *Familiar Quotations*: I don't believe so. There is nothing in the Chinese literature that corroborates that notion. At least nothing attributable.

Including Mandarin, Cantonese, and local dialects, all Chinese is written in the Han alphabet. Unlike the phonetic alphabets of the Western world, the Han is a pictorial alphabet. Therein, each character was designed as a miniature brush-line-drawing, symbolically representing the word (or part) it means to convey. Add-ons to a basic character will alter its sense of meaning or sound. A single character can mean several different things, depending upon the context of the rest of the sentence surrounding it. Conversely, in some cases there is more than one word that you can use to express a single thought, especially if the new character appears more aesthetic within its surroundings.

For instance, the phonetic sound "Dan," a male name in several world cultures, is written 旦. That symbol, depicting the sun above the horizon, could mean the hour of dawn (daybreak); or egg; or bright boy (all ostensibly from the same root idea of a beginning). Any one of those meanings depends upon the flow of the rest of the words in the writer's thought. That symbol cannot be misconstrued for the sun headed down toward the horizon (sunset) because the horizon line symbol would be drawn above the sun; there is no such picture-word.

In his late third-century AD classic tome, *Wen Fu* (*The Art of Letters*), Lu Chi (Ji) wrote, "The character you choose within any sentence should meet (or approach) these three criteria; it should have: the correct sense of meaning; appear on the page as the most pleasing to the eye; and, cause the most pleasing sound when read aloud." It must have been difficult to have been a successful author there, then. Perhaps that's why so much from that good era still exists.

It seems hardly possible that that advice could be a Chinese proverb because of the nature of Han words. It would be redundant and un-Chinese to write, "One picture is worth more than ten thousand pictures."

I. S. Turgenev, in 1862, penned *Fathers and Sons*, in which he takes a broader and more nebulous view with his derivative line, "A picture shows me at a glance what it takes dozens of pages of a book to expound" (chap. 8). Perhaps some writers in old Russia didn't do all that well at deadhead deletion.

Use photographs, schematics, sketches, charts, tables, graphs, etc., liberally to augment, not to replace (in addition to, not instead of), the written word. Always draft your composed description, then refer to the related graphic if there is one. In every graphic caption, include the related text-index number. This allows you (if layout space demands) freedom to put a graphic on a different page from its related text. In a narrative description, never write a statement like, "Arrange switches on control panel as shown in photo 6.53." Describe the switch arrangement completely in the text, then refer to the photo.

5.7 CAPTIONS

A caption should make the graphic more explicit. Some captions are so complete that you can envision the graphic without actually seeing it.

Shelter, canvas, to cover aircraft maintenance one-man work area; lightweight, portable; assembles quickly.

That tells you what it is, its purpose, its materials, and its advantage.

Daylight aerial photo of crash site, 4.7. Shows direction, skid tracks, and vehicle wreckage, plus position on sharp curve and wreckage of stalled tractor at impact point. Also shows total absence of warning signs or lights.

That caption calls attention to important elements and their related positions in a broad area photo.

Jefferson University 1922 freshman class, picture 6.3. Future president, Thor Brockway, is in third row, fourth from left.

It pinpoints focus subject in a large, identified group.

New model, miniature GPA dash-mounted device, 3.25. U.S. penny included to show relative scale.

Caption gives all necessary information to augment the text.

5.8 DESCRIPTION WRITING METHOD

Use as many of these that apply, not necessarily all of them.

5.8.1

Name the entity. First tell your reader what your focus subject is.

5.8.2

Tell what it does, or its uses.

5.8.3

Provide general characteristics: shape and condition.

5.8.4

Provide physical characteristics: length, width, height, name and number of principal parts, and special features.

5.8.5

Provide special characteristics: material, color, finish, and electrical or mechanical specifications.

5.8.6

Provide additional identifying information as needed: catalog listing with stock number, developer's name, and trade term (e.g., Geiger counter, Surge milker, Norden bombsight, Benkleman beam).

5.9 DELIVERY

Write a narrative, not a list of facts.

5.9.1

The government way may appear clumsy, but it is the most explicit, for example, "1 ea. Desk, wood, brown" is preferable to "1 ea. Brown wooden desk." The subject is not a *brown*, it is a *desk*.

5.9.2

Use *factual* exposition. Impressions are opinions; do not offer opinions in descriptions unless requested in the charge. Be precise.

5.9.3

Avoid qualifiers; they weaken the statements. For example, "It is pretty light" is nebulous, "It is light" is precise. Same goes for "very."

YOUR WORD PICTURE MUST CREATE A
REALISTIC IMAGE IN YOUR RECEIVER'S MIND.

—A.L. Morris

6
COMPARISONS

6.0

In writing a technical letter, an order, or a magazine article; for example, about a concrete housing for a synchrotron in a tunnel, you might describe it as *"a buried concrete closed circular tube with a 10-foot-diameter vertical cross section. The locus of cross-section center points describes a 2,000 foot circumference horizontal circle, the tube's centerline . . . "* plus more details of course. If you have preceded that description with "a doughnut-shaped structure" or "a torus" (depending on your real reader levels), how much easier it would have been for him to envision the tube. That added simile statement is a *comparison*. While a description tells what something *is*, a comparison tells what something *is like*. Ergo, a comparison becomes a principal aid in enhancing a description to make a better reader image and to emphasize a point.

The *Consumers Union Reports* are consistently excellent examples of the maximum use of comparisons to validate their efforts.

While a comparison shows some form of similarity, a *contrast* shows differences or departures. Strunk and White (*The Elements of Style*, 14th ed.) state that the difference between writing compared *to* and compared *with* is a matter of time frame. The *with* indicates the entities are contemporary *with* each other; the *to* indicates they are of different periods, then *to* now.

Another frequent mistake is in writing or speaking "different *than*." Correctly stated it should be "different *from*." A difference is a departure *from*.

6.1

Guided by both your chosen reason(s) for writing and your target reader(s), your comparisons will tell him the *relative* merits among *similar* or *like* entities with regard to each other or with regard to a fixed standard. In describing a wholly new thing to a reader, liken it to a similar thing he knows to help him evolve his new image: the same logic applies to ideas and concepts. Use charts, graphs, or spreadsheets (with explanatory text) to display comparisons of nondepictive items. In photos, always include a recognizable item that will show the relative size of the subject, for example, a coin among small things or a bicycle among machines.

6.2

You can generate comparisons *only* within these four constraints:

6.2.1

Comparisons must concern only *similar* or *like* entities. But be careful about the limit of your umbrella of affinity; it cannot be too broad. Both a structural compression member and a razor blade are made of steel, but they are comparable neither in size, mass, nor use. You cannot compare horses and peaches, saying they both are living things. Parts of biology is far too broad a reasonable spectrum. Although once, a young man in a short course in Chicago flippantly claimed, "You could, if you'd regard them both as things you could steal!"

6.2.2

Make comparisons from the same point of view. You must regard all considered compared things from one position. Things appear as different sizes with regard to each other when each is perceived from a different viewpoint. On a freshman engineering field trip to Boulder Dam, I stood at its toe and looked straight up some 700 feet to its top. To me, then, Boulder Dam was the biggest thing I had ever seen. Fifteen years later, on a trip from New York to Los Angeles, I saw it from my aircraft window, 5000 feet above. Boulder Dam appeared the same size as my thumbnail. It hadn't changed its dimensions, but my viewpoint had. Matthew 19:24 suggests you can't pass a camel "through the eye of a needle." You *can* if you perceive the camel through the eye of the needle you are holding 100 feet distant from the beast.

6.2.3

Make comparisons under the same conditions. Two grades of diesel fuel may behave identically in long-run buses, but one is not efficient for the start-and-stop traffic of a New York City local bus.

6.2.4

Make comparisons within the same environment. Both tungsten wire and an alloy steel wire may be suitable for some tasks in our normal atmosphere, but their behavior in a glass-enclosed vacuum is different.

And comparisons must be balanced. Statements are useless and unacceptable if they read like this: "It's warmer in Ithaca than it is in the summer." Or worse, "Compared to Norway, living is more expensive."

6.3 DELIVERY

You can achieve the most from a comparison by the creativity and strength of your presentation. And do not take that statement to include creativity in dealing with the facts. Facts are not tamperable. Choose your words skillfully to make your similarities appear closer or emphasize a greater separation in your contrasts. An easy way is by a parallel chart. Make each subject a column heading. List your considerations down the left-hand column, preferably, but not necessarily, in order of importance. You have seen in sections 4.1 and 4.2 how easy that is to do. Always keep the first line of every listed consideration horizontally on the same line in all columns, leaving a blank space where one entry is shorter than the next. Failure to do that leaves a comparison unaligned, and leads to confusion.

Make your entries razor sharp by including precise details, but don't lose your thought line in a thicket of words. Steven P. Waterman, Esq., then law clerk in the Ithaca court system, made a chart comparing the codes of ethics of the legal profession vis-à-vis the public accounting profession.

He set up his chart (fig. 6.3.1) as just suggested, but altered it to include all the *necessary* details while allowing the principal considerations to tell their uncamouflaged story. He copiously footnoted all tangential details, references, and minutiae on a separate sheet (fig. 6.3.2) to make them readily available, but he kept them from cluttering the main thrust.

Always write a summary paragraph at the end of the chart to steer the reader to your intended conclusion. Figures 6.3.1 and 6.3.2 show a reader-friendly comparison statement between the two professions. Imagine how difficult that point would have been to make had all the footnotes been left in the text.

6.4 NARRATIVE PRESENTATION ALTERNATIVE

W. LeDuc of the U.S. Coast Guard had to investigate and report on a comparison between two electronic systems, to go to both administrative and to

FIGURE 6.3.1. LEGAL VS. ACCOUNTING CODES OF ETHICS

LEGAL CODE OF ETHICS	ACCOUNTING CODE OF ETHICS
Primary duty is to the client; secondary duty is to the courts and the public.[1]	Primary duty is to the public; secondary duty is to the client.[1a]
Absolute duty to preserve the confidences and secrets of the client.[2]	Absolute duty fairly and fully to present to the investing public information received from the client.[2a]
Must at all times act in the best interest of the client.[3]	Must at all times remain independent of clients' interests.[3a]
Must not refuse services to a client solely by reason of latter's inability to pay.[4]	Must at all times receive payment for services to avoid allegations that independence is lacking.[4a]
Has a duty to counsel the client concerning nonlegal aspects of decisions that influence the legal aspects.[5]	Has a duty to avoid counseling his client concerning nonaccounting issues.[5a]
Must represent a client zealously within the bounds of the law.[6]	Must represent the client objectively, resolving all doubts in favor of full disclosure to the public.[6a]
Attorney may advise a client concerning future conduct when requested.[7]	Accountant may not attest to the accuracy of a client's forecasts.[7a]
Attorney has duty of continuing representation to client who refuses attorney's advice on a material issue, provided the attorney's ability zealously to represent is not prejudiced.[8]	Accountant must withdraw from employment if a client refuses advice on a material issue.[8a]
Attorney may not practice public accounting while simultaneously practicing law.[9]	An accountant may simultaneously practice law while working as a public accountant.[9a]
An attorney's violation of an ethical consideration is not a per se ground for discipline.	An accountant's violation of an ethical standard is a per se ground for discipline.
Ethical canons and considerations provide guidelines to which all attorneys should aspire.	Ethical rules provide minimum standards for all members of the profession.
Ethical violations requiring discipline are uniformly enforced among the authoritative bodies of the legal profession.	Ethical standards are not uniformly enforced among the authoritative bodies of the accounting profession (AICPA, state societies and state boards of accountancy).[10]

The Legal Code of Ethics portrays the attorney as the agent of the client to whom undivided loyalty is owed; the Accounting Code perceives the accountant as providing a necessary service to the client, but having a supervening duty of full disclosure to the investing public.[11] The attorney aspires to provide a complete advisory service, encompassing non-legal and legal decisions, while the accountant solicits only that amount of information relevant to financial disclosure. Ethical considerations for attorneys are goals to strive for, for accountants they are the minimum required. (Footnotes on next page.)

Steven P. Waterman, Esq., Law Clerk to Judge Dean, Ithaca, NY

FIGURE 6.3.2. FOOTNOTES TO FIGURE 6.3.1

NOTES

1. American Bar Association. Special Committee on Evaluation of Ethical Standards, *Code of Professional Responsibility* (Chicago: American Bar Association, 1969) (hereinafter *CPR*). "Ethical Considerations" (one of the three main parts of *CPR*; hereinafter EC) 5 · 1. See generally, Thomas D. Morgan and Ronald D. Rotunda, *Professional Responsibility* (Mineola, NY: Foundation Press, 1976). See also, footnote 11, *infra*.
1a. American Institute of Certified Public Accountants (AICPA), *Code of Professional Conduct* (hereinafter *CPC*), Rule 1 (Independence) and Rule 301 (Confidential Client Information). See also, footnote 11, *infra*.
2. *CPR*, EC 4 · 1; see also, Disciplinary Rule (DR) 4 · 101.
2a. *CPC* Rule 202 (Auditing Standards) and Rule 203 (Accounting Principles).
3. *CPR*, EC 5 · 21.
3a. *CPC* Rule 101, *supra*.
4. *CPR*, EC 2 · 16 and 2 · 25.
4a. *CPC* Rule 302 (Contingent Fees).
5. *CPR*, EC 7 · 8.
5a. *CPC* Rule 201 (D) (General Standards—Sufficient Relevant Data).
6. *CPR*, Ethical Canon 7, EC 7 · 1 to 7 · 39; see also, DR 7 · 101 and 7 · 102.
6a. *CPC* Rule 102 (Integrity and Objectivity)—only exception is taxes.
7. *CPR*, EC 7 · 8.
7a. *CPC* Rule 204 (Forecasts).
8. *CPR*, EC 2 · 32 and 7 · 5.
8a. *CPC* Rules 202 and 203, *supra*.
9. DR 2 · 102 (E).
9a. *CPC* Rule 504 (Incompatible Occupations), amended recently to state: "A member who is engaged in the practice of public accounting shall not concurrently engage in any business or occupation which *would create a conflict of interest* in rendering professional services" (emphasis added). The previous version had "impairs his objectivity" in place of the above italics, with the added phrase, "or serves as a feeder to his practice," at the end. Obviously under the old version, law would have provided a feeder, whereas under the new amendment, law will not create a conflict of interest per se.
10. Roger H. Hermanson, Robert H. Strawser, John M. Saada, and Stephen E. Loeb, *Auditing Theory and Practice*, rev ed. (Homewood, IL: Richard D. Irwin, 1980).
11. "In *In Re American Finance Co.*, 40 SEC 1043 (1962), the SEC contrasted the lawyer's duty with the independent accountant's duty as follows: 'Though owing a public responsibility, an attorney acting as the client's advisor, defender, advocate and confidant enters into a personal relationship in which his principal concern is with the interests and rights of his client. The requirement of the Act of certification by an independent accountant, on the other hand, is intended to secure for the benefit of public investors the detached objectivity of a disinterested person. The certifying accountant must be one who is connected with the business or its management and who does not have any relationship that might affect the independence which at times may require him to voice public criticisms of his client's accounting practices.'" Arthur F. Mathews, "Effective Defense of SEC Investigations: Laying the Foundation For Successful Disposition of Subsequent Civil Administrative and Criminal Proceedings," 24 Emory LJ. 567, at 628, 629, n212 (1975).

FIGURE 6.4.2. COMPARISON PROTOTYPE: NARRATIVE FORM

(For Electronic Supervisory Audience)

Comparison of electronics installation, change, and maintenance (EICAM) with the proposed electronics maintenance teleprocessing network (TELENET)

EICAM	TELENET
Eight different ADP forms	One message format
Surface mail, then enter	Direct to computer via modem
No individual access to data	May accept any terminal input upon demand
Two weeks to obtain reports	May be accessed to produce reports at any time
Between 10% and 20% entry errors	4% input error estimated, but CPU may be programmed for automatic correction
Lowest computer priority	Only computer priority
6 weeks to 4 months to modify software	48 hours maximum to modify and install software
Averages 34% reporting accuracy	Estimated 99.8% reporting accuracy
Maintenance requires 6 man-years effort	Maintenance estimated 2.4 man-years annually
Two weeks to update master catalog	Four hours to update master catalog
Requires 16 cubic feet backup storage	Requires 2 cubic feet backup storage for resident and database discs
Special Reports obtained in two days	Special reports obtained at will
Inventory updated semiannually	Inventory updated upon demand at any terminal
No backup reporting system	Disc storage and TTY bypass backup
No direct access except phone	Screen mail plus phone feedback
No additional hardware required	32 modems and 12 TTY interfaces required

TELENET provides the prime input data source, field units, and District Offices, with direct access to the information. Being able to use the data input will generate a greater interest in the accuracy of those data.

    The basic difference between the two systems is the responsiveness to the needs of the Organizational and Intermediate level activities and acquisition cost. The TELENET system may be used by any cognizant activity at any time for pertinent electronics maintenance data. Response will be instantaneous and accurate. The initial cost of TELENET installation is estimated at $240.6K; this amount may be decreased by the annual maintenance cost for EICAM, leaving a first-year savings of $86K.

W. LeDuc, U.S.C.G.

(continued on next page)

COMPARISONS   69

(For Non-electronic Supervisory Audience)

Comparison of electronics installation, change, and maintenance (EICAM) with the proposed electronics maintenance teleprocessing network (TELENET)

The EICAM system: eight ADP forms; depends on surface mail; cannot accept individual data access; requires minimum two weeks to deliver reports; subject to 10%-20% entry errors; has lowest computer priority; requires from six weeks to four months to modify software; averages 34% reporting accuracy; requires six man-years effort annually; takes two weeks to add new equipment type to the master catalog; requires 16 cubic feet of backup storage; search reports may be obtained in two days; inventory has no direct access from the field units; uses existing computer resources.

The proposed TELENET system: requires one message format; depends on data link and interface modems; may accept access upon demand; may be accessed and produce reports at will; subject to estimated 4% input error, but CPU readily may be programmed for automatic correction; EPM is TELENETs only priority; requires 48 hours to modify and install new software; will average an estimated 99.8% reporting accuracy; requires 2.4 man-years effort annually; requires two cubic feet of backup storage for resident software and database discs; search reports may be obtained in 2-5 minutes; inventory updates run upon demand at each microprocessor; has 32Kb input buffer storage and TTY input access-loop may bypass fault or store input data until hardware error corrected; field units may submit direct feedback via screen mail option; will require standardization of existing microprocessors and installation of all interface equipment.

The basic different between the two systems is the responsiveness to the needs of the Organizational and Intermediate level activities and acquisition cost. The TELENET system may be used by any cognizant activity at any time for pertinent electronics maintenance data. Response will be instantaneous and accurate. The initial cost of TELENET installation is estimated at $240.6K; this amount may be decreased by the annual maintenance cost for EICAM, leaving a first-year savings of $86K.

W. LeDuc, U.S.C.G.

FIGURE 6.5. GRID COMPARISON CHART: MAJOR ROCK FORMATIONS IN MANHATTAN

	MANHATTAN AND HARTLAND SCHISTS	INWOOD MARBLE	FORDHAM GNEISS	GRANITE PEGMATITE
FOLIATION	Moderate to well foliated	Not foliated	Poorly to well foliated	Not foliated, intrudes pre-existing rocks
FOLDING	Layers gently to moderately folded	Not applicable	Layers intensely folded	Cuts across or between foliation
CLEAVAGE	Breaks along mica layers	Breaks irregularly	Breaks irregularly	Breaks irregularly
TEXTURE	Medium to coarse grained	Medium to coarse grained	Coarse grained	Coarse to extremely coarse grained
COLOR	Gray and brown to black	White to blue-gray	Gray and brown to black	Gray, white and pink
WEATHERING	Weathers to a rusty brown to black; mica layers weather more readily	Weathers to a rusty brown; rock crumbles easily into sand grains	Weathers to a dark gray to black	Weathers to a gray to black
MINERALS	Quartz and feldspar layers, mica layers garnet, kyanite, sillimanite, magnetite, staurolite, tourmaline	Calcite and dolomite, mica, pyrite, tremolite, graphite	Quartz and feldspar layers, mica, feldspar and hornblende layers	Quartz, feldspar, mica, garnet, beryl, tourmaline, spodumene

CHERYL J. MOSS—MUESER RUTLEDGE CONSULTING ENGINEERS

technical management. He compiled the results of his investigation into the best format for each of his addressees. His comparison chart with its closing summary paragraph (fig. 6.4.1) makes available all the necessary information for the technical audience. The narrative version (fig. 6.4.2) provides the identical information in a comprehensive, compact form, including the *same* summary conclusion.

To write the narrative presentation, do a short introductory paragraph. Then, taking each entry in the first column (in its order of importance), restate each in a proper English sentence, followed by the next item, to make a paragraph. Then repeat the method for the second column, keeping the characteristic sentences in the same order as the first. And so on for each additional listing. Close with the same summary paragraph as was in the chart display.

6.5

To improve the good rapport between their engineers and the many Manhattan foundation contractors with whom they deal daily, Cheryl J. Moss, a geologist with Mueser Rutledge Consulting Engineers, developed an easy-to-read, comprehensive comparison chart (fig. 6.5) of the four principal subsurface rock types with which they all had to work.

Posting those types as four column heads, she listed the seven constant characteristics among them down the left-hand column. The usable work information, clearly and briefly stated in the grid, enabled everyone involved to be at the same place, and thus do the job efficiently.

6.6

Circa 1625, the Rev. John Donne reiterated, "comparisons are odious." That fragment has become a cliché that somehow casts a pall over the notion of comparisons, as if they have "fallen to the dark side." In *De Lavdibus Legum Angliae*, a century and a half earlier, John Fortescue wrote the original bromide, "comparisons are odious," with a different intent. To the contrary, comparison is a useful tool in technical and administrative writing.

> A GOOD LIKENESS WILL TELL A READER
> HOW SOMETHING LOOKS OR WORKS,
> AND WHAT IT MAY MEAN, OR BE WORTH.

7
MOTION

7.0 MOTION

Motion is the third of the three elements comprising the material substance dimension of a communication. The other two elements, *description* (chap. 5) and *comparison* (chap. 6), are *static*, and if properly written should give your reader a clear impression of a stationary subject. This chapter will tell you how to transmit in writing the notion of motion, movement, dynamics.

Principally motion is not an isolated writing tool, although it may in some instances be used alone. More likely you write motion seamlessly in conjunction with either or both description and comparison, each a component part of the material substance.

You write motion to convey instructions, directions, recipes, musical scores, prescriptions, progress reports, shop orders, work orders, change orders, and manuals; but not specifications. Specification writing is whole subject unto itself, incorporating all the basic principles contained in this volume, then extrapolating them to suit the *specific* requirements of your document.

Your target reader (of any of those items just cited) *needs* the document information it carries, thus he becomes a captive audience. Yet you must maintain his attention throughout for him to receive your complete, detailed information.

7.1 CONCEPT

Before digital cameras and video, all motion pictures were made on cellulose acetate–based film. Consider, for instance, watching a football film projected on a screen. You'd see Tom Brady's right hand holding the ball; he cocks his arm, then flings the ball spiraling toward the receiver, Randy Moss. After the flick ends and the lights come on, you hold the film strip in your hands. There is nothing moving in the film. You see a series of consecutive still frames, each one showing Brady's arm for an instant in a different position forward of the preceding frame position. In calculus the symbol dx/dt indicates a change of position with regard to time. When the still frames are run rapidly in consecutive order, you get the impression of motion. Thomas Alva Edison (the incandescent light bulb) materialized that phenomenon and evolved mechanical motion picture devices both for filming and projecting. But Eadweard Muybridge,[1] a late nineteenth-century photographer, set up a long straight row of 24 string-tripped still cameras. Each shot a photo in rapid succession after its predecessor while a horse galloped past them. From observing the change of position of the horse's hooves as the time and pictures progressed, he proved that at one point in each stride all four of a horse's feet are off the ground, until then a long-time point of conjecture. In 1875 Muybridge published three photo-illustrated books: *The Horse in Motion*, *Animals in Motion*, and *The Human Figure in Motion*; the last includes his classic photo series, "Girl Kicking a Hat." Consider Muybridge's 1875 publication date beside Edison's *kinescope*, that made its debut public appearance on April 14, 1894.

That graphic background provides for you the basis for writing motion. Instead of consecutive still pictures showing changes over time, you write a series of sequential statements in close consecutive order; the resultant sequential readout will convey the notion of motion.

7.2 INSTRUCTIONS

Instructions are the most frequent application of motion writing. Telling a reader, clearly, how to do something (possibly entirely new to him, or worse, something he thinks he knows all of everything about) is not easy. Again it brings up both your reader's knowledge datum and his reader comprehension level. There is no one-size-fits-all instruction for writing instructions; but I do offer an umbrella method, sufficiently broad that you may use what you need of it, and leave the rest.

Caveat: Usually it is not prudent to write an instruction for a task of which you have no personal experience. Of course, that is not always possible, but if you've done it yourself, you know firsthand of what you are writing. Thus, you pass it on more confidently, and the reader uses it more confidently, with full trust. An old fragment from the golden age of Chinese proverbs sagaciously advises, "To know the road ahead, ask those coming back."

7.3 INSTRUCTION PROTOTYPE

To give you the most from this exercise, I've confected a multifaceted operation that includes as many separate situations as reasonably possible. The example shall be an instruction for putting together two separate assemblies from preselected fabricated parts. The two assemblies shall, in turn, be combined into one single finished installation for a larger machine.

7.3.1 FIRST STAGE.

7.3.11

Cite the installation, and the specific first stage.

7.3.12

State what you wish to accomplish in this stage, and why. The operator will then know where this is going.

7.3.13

State this caveat verbatim. It seems obvious, but should not be left unsaid: "Please read *all* instructions before you do anything else." Too many clever, impatient people dive right in, doing it while they read until they come to an abrupt stop because they couldn't anticipate the next condition. Better to be aware, as just stated in the section 7.2 caveat.

7.3.14

List a bill of materials (B/M) for all stages, through to complete installation, to avoid the reader's having to pause the action while he runs out to get a part. Include everything needed for a complete, uninterrupted operation, then restate the items needed *only* for the first stage.

7.3.15

List all cautions and warnings for the safety of the operator and nearby persons. This is for the first stage only; each stage will have its own hazards in that stage at hand. Safety is the top priority. Each stage should be treated in its own time (app. 7.3.15).

7.3.16

List boundary conditions, limits, and parameters for operation in the first stage only.

7.3.17

Procedure: the actual step-by-step process in chronological order to complete the first assembly using all the parts listed in the B/M for first stage only (app. 7.3.171 and 7.3.172).

7.3.18

List all available spot checks for use along the march in stage 1 only, to be sure the operator is in line to the end result.

You are done with the first assembly.

7.3.2 SECOND STAGE.

7.3.21

Restate the overall title and the specific subtitle for stage 2.

7.3.22

Cite the objective for stage 2 only.

7.3.23

No need at this point to restate the caution caveat.

7.3.24

For operator comfort, restate that portion of the B/M and special preps for the second stage only.

7.3.25

Carefully state specific new safety cautions for the second stage only.

7.3.26

Cite whatever additional limits and parameters are required in the second stage only.

7.3.27

State in detail all the steps, in turn, required to complete the second assembly using all the materials listed in section 7.3.24.

7.3.28

Cite all available spot checks along the line for stage 2 to be safely on track. You are done with the second assembly.

7.3.3 FURTHER STAGES

If there are three or more assemblies, you repeat the protocol for stage 2 for each unit for as many component assemblies as needed.

7.3.4 INSTALLATION BUILDING

Combine all the assemblies into the installation after all of each stage is completed. The instructions for the combining stage should include:

7.3.41

Appropriate title.

7.3.42

Accomplishment goal for the full installation.

7.3.43

Caveat not needed at this point.

7.3.44

All things necessary to finish the entire job.

7.3.45

All new safety precautions.

7.3.46

All new limits and parameters.

7.3.47

Details of all procedures required to finish the job.

7.3.48

All spot checks. You should now have a complete working installation.

7.3.5 FINISHING

To put this installation into the big machine, follow all the steps in section 7.3.1.

7.4 THE LOGICAL, FOURTH CONSIDERATION

I have skimmed through some thirty-odd books on writing on the shelves of one technical and one major chain bookstore in New York. Their overall consensus on writing instructions comprised three principal points. All instruction, they say, must be *understandable, concise*, and *grammatically correct*.

Do Not Read This.

That terse instruction is clearly comprehensible; its four words are in their closest compass (eliminating any one word will alter or lose the sense of meaning of the instruction); and, as it stands, the statement is grammatically correct. But it is an instruction impossible to follow because it is an absurdity. Thus, while agreeing with the consensus, I must add a necessary fourth requirement. Yes, an instruction must be understandable, concise, and grammatically correct, but it also must be *logical* (app. 7.4).

7.5

This entire volume is dedicated to professional expressions in English. Although I am passably proficient in only one additional ethnic idiom, I make no presumption about conveying ideas in any other language beyond English.

Some specific professional practices have universal parlances peculiar to them; symbols that override national tongues. Consider: Two currently contemporary violinists, a Brazilian and a Hungarian, can play a perfect duet, composed centuries earlier by an Italian. None of those three would be able to speak outside his native tongue, but the fiddlers can play in compatible synchronicity because the composer wrote his instrumental instructions in the language universally understood by all musicians. That is, black dots, some with flags, maybe an octothorpe, and a few other symbols, all on a score sheet of staves (app. 7.5), each of five lines enclosing four spaces, plus additional ledger lines above and below when needed. From that sort of written message, musicians everywhere in the Western world can play a melody to sound exactly as its originator had envisioned, then instructed.

And that performance is only the *first* half of the train of thought conveyance. The violin duo converted the composer's visually written musical instructions to audible musical sounds. Those specific sounds, in turn, transmitted in air, created in the listener's mind perhaps the image of a bird in flight: the exact image the composer had in mind as he wrote in another language, in another era, half a world away.

Again the same *dx*, the single frames of a film strip run in rapid succession, to transmit the *dt* notion of motion. In his description of the Virginia countryside (sec. 5.3), Richard L. Klein moves you with the flow of the James River as it moves from its source to its outfall at Hampton Roads. While reading that, in my mind's ear I heard Bedrich Smetana's tone poem, the *Moldau*, also about a river, flowing through the Bohemian countryside to the sea.

Similarly, a Venezuelan physician writes an instruction to a Bulgarian pharmacist to compound a specific remedy for dyspepsia, in the language of the apothecaries. Prescriptions ("scripts") begin with the traditional figure, ℞. That symbol evolved from the Latin *composition*, via the Italian *recitta* (*recipe*), meaning *to take*. That, in the sense of instructing the druggist what ingredients *to take* to put into his *compound* to yield the desired *composite*. The prescription, ℞, then goes on to instruct the patient *how to take* the compound to alleviate his ailment; "just what the doctor ordered." Toward universal common understanding, Linnaeus[2] organized and classified all things botanical into Latin, so everyone envisioned the same thing.

Although the predominant part of this *motion* writing chapter deals with composing instructions, do not lose sight of the *principal* use for this entire element of the main idea, *thought transference*. Descriptions and comparisons tell your reader about *statics*. By using this important element, *motion*, you tell him how to move entities toward your intended goal.

> EVERY BODY CONTINUES IN ITS STATE OF . . .
> UNIFORM MOTION IN A RIGHT LINE.[3]
>
> —Sir Isaac Newton

NOTES

1. b. Edward James Muggeridge (1830–78).
2. Karl von Linné, Sweden (1707–78).
3. Sir Issac Newton, preface to *Philosophiae Naturalis Principia Mathematica* (London: Samuel Pepys Presses, 1687). A body in motion tends to remain in motion in a straight line.

8
SHORT FORMS

8.0

A few years ago the late Brother B. Austin Barry of the Manhattan College Civil Engineering Department sent me an eight-page, 6" × 8" slick paper pamphlet. Each page was packed to its narrow margins with what my printer calls agate type (the smallest size font a normal eye can read without a magnifying glass). It was, in its second revision no less, the *constitution and bylaws* of a small engineering organization. I cite verbatim Brother Austin's transmitting buck slip:

Comments:

(1) I've not read this.
(2) Neither will you.
(3) Nor will anyone else.
(4) No one will follow it correctly (by reason of no. 3).
(5) I don't care.
(6) Neither will you.
(7) It's too long.

/s/ Commentator

In his humorously terse message, Brother Austin summarized the exact antithesis of the pamphlet. His well-taken point to that organization, and to me, was, at either of the two revisions, had the editors taken the effort to condense the document, it would have been read, understood, and followed. Thereby it would have been useful to its intended audience. In its present unabridged form, it is not any of those things: "It's too long."

8.1 USES

A short form is the product of a specific process that shrinks or abbreviates, but does not truncate, a finished, fully written original text. It includes no additional material, nothing new that does not appear in the long, written document. In effect, a short form is a carefully miniaturized version of the original document, condensed to fit a specific need. It retains a few details germane to the *purpose* (reason) for that unique need.

8.1.1 ABSTRACT

An *abstract* appears at the top, preceding a finished document, but is written only *after* the entire document is complete. It should tell a prospective reader what the narrative is about, but *not* tell the narrative. Tell what *kind* of information is in the paper, but do not give the information itself. A researcher won't read your whole document if he can find what he seeks in the abstract.

8.1.2 INFORMATION RETRIEVAL

You may put your abstract, as is, on the Internet to enable a researcher to determine at a glance if its full document contains the information he seeks.

8.1.3 LETTER REPORTS

See Executive Summary (sec. 8.5).

8.1.4 PRESS RELEASES

State the key point in the opening sentence, name names, include a quote, then follow, in short, the procedure for writing an Executive Summary (sec. 8.5).

8.2

On two occasions ten years apart, to get a better feeling for how I taught the short-form exercise, I spent a day each time at two major Manhattan bookseller's technical writing sections. There, I ruminated over more than a dozen titles. Although none referred to the subject as "short form," many dealt with it solely as "abstract" writing. They all told me things like what an abstract is, what it might (or might not) include, and where it belongs in a document. Not one came to grips with a method for producing one. To be of real value to any writer, this volume does that, now.

In chemistry at New York's Stuyvesant High School, I learned that to replicate a substance (or variations) in any quantity, you must, by analysis, isolate and identify its every active element, then by synthesis recombine them. If the result is a rejuvenated original substance—success!

Using that logic, I listed 13 words I know to be of the short-form genre:

abridgment	epitome
abstract	essence
condensation	précis
conspectus	quiddity
digest	résumé
distillation	summary
	synopsis

Then I looked at each word in three different dictionaries (Merriam-Webster, Random House College, and The Reader's Digest Great Encyclopedia). Noting the 39 definitions, each on a file card, I laid them out and sought common lines among them. Not every one included the same elements; many had some, a few had them all. The four common threads to appear most often defined the *characteristics* of a short form. They are:

8.2.1 REDUCE *PRINCIPAL MATTERS* TO *FEWEST WORDS*.

Eliminate minutiae and embellishments. A young man learning to play the tuba did his first solo. It sounded like a Brazilian Samba in triple time. The teacher stopped him, asking what he was playing. His reply, "The score," prompted the teacher to have another look. Then he admonished the novice, "Please, play only the notes, never mind the fly specks" (app. 8.2.1).

8.2.2 GIVE A *GENERAL VIEW* OF THE *OVERALL SITUATION*.

That is perspicacity, the skill to envision a perspective view of the *big picture*. A standard 8"× 8" print of an aerial photo shot at 5,000 feet will show all of Central Park, shrunk to scale (short form). An 8" × 8" sod block cut from park ground shows only 64 sq. in. of sod (truncation).

8.2.3 SEPARATE CAUSE FROM EFFECT, AND LABEL EACH.

In some instances, what is a cause can become an effect in the same paragraph.

Cause	*Effect*
change of ambiance	precipitation
precipitation	soil flooding
soil flooding	particle movement

particle movement
differential settlement
foundation displacement

differential settlement
foundation displacement
structural damage

8.2.4 RETAIN A SENSE OF BOTH *SUBSTANCE* AND *QUALITY* (SEC. 8.6).

Of the 13 words in the section 8.2 list, only one is unique, *quiddity*. It differs from all the rest because it specifies both substance and quality as requirements. You must have substance to have a written effort. Quality is the nonquantifiable attribute that raises your document above others, by making it more enjoyable reading. That does not mean simplistic language, nor flowery words.

A top-line electronics company received a request for proposal (RFP). It had no item in inventory then that met the specifications. Also, the three in-house engineers qualified to design such a piece were all gainfully occupied completing a contract whose due date was well past the proposal return time. The firm opted to decline, but had to do so in writing to maintain its place on the source list. It demurred with this e-mail:

No item. No chance. No bid.

That message had been boiled down to what one agency recently called "an irreducible minimum." You might say it could have eliminated two of the triple, redundant "nos" and been briefer still. But the reduced *No item, chance, bid* is not as precise, leaves room for misinterpretation, and lacks the *quality* element just noted. Do not cut messages to where you lose the thought. At a market, one vendor put up a sign in large letters proclaiming, FRESH FISH SOLD HERE. The first customer suggested he didn't need the "here" because both he and the vendor were indeed here. Thus he lopped it off and left FRESH FISH SOLD. The second customer thought the "sold" was useless because that was the only reason they were both there. The sign was reduced to FRESH FISH. The next buyer said he would not patronize a monger whose fish were not fresh, thus because the wares were only so, the "fresh" came off, leaving just FISH. That was removed by the vendor because people could smell them.

Don't brief your message out of existence. You must leave sufficient wordage to convey your thought, depending upon the reader, then you may add a few quality words to make it readably appetizing. In-house, depending upon the closeness of the group, you may make some reasonable assumptions, but the principal matters must be what remain.

In summer, a hotel restaurant places two urns in the kitchen, one for iced coffee, the other for iced tea. Not all the servers know the iced coffee is always in the left urn, thus they must be labeled. Because the audience is small and experienced, and both are iced, only one urn need be identified to know what both are. Then the "ea" could be cut from "tea," leaving only the letter "T" on the right urn to explain the whole display to a T.

Contrast that efficient solution (to fit the target audience) to the previous fish story. Once you've mastered the way to write short forms, then your reasonable judgment grows to be a governing factor in your clear communication.

8.3 PROCESS

This six-step method will work whether you do an abstract of a writing of your own, or if asked by a colleague or boss to write an abstract of someone else's writing.

In a class at Army Munitions Command, Betty Zweig was asked to read a published article, then write a synopsis of it. She encapsulates the essence and quality of the article facts, without details.

> Extensive retraining is unnecessary to produce straightforward technical communications. The process is simple: Know your subject; Break it down to the essentials; Write it in clear, concise, correct English.
>
> No progress comes from confusing the reader. The subject may be scientific, but the idiom is still English, and English is the vehicle by which the lay-reader or the highly specialized scientist hopes to exchange his messages.

8.3.1

Read the subject paper thoroughly. Skimming won't do here. Then ask yourself this single question, "What does this document try to tell me?" If after the reading you are not absolutely certain what the document and its implications are about, do not undertake a short form. Without that definitive understanding, you could lead to confusion or create error. Now read this abstract taken from a study report. Obviously not fully grasping the subject, the short-form writer dances around his subject, drowning it in 144 words.

> The study of Human Tolerance to Random Vibration was separated into two pilot studies and a main experiment. The analysis of the

pilot study data indicates that subjective judgments of ride severity may be measured with precision and that they may display a relatively simple relation to the characteristics of the ride. For the main experiment, the subjects were consistent in their ratings of ride severity. The difference in total power (gain) and in frequency composition accounted for most of the variation due to experiment conditions. The attempt to give numerical methods of predicting vibrational severity ratings did not meet with good success though useful results were obtained. The indications from the results of this experiment are that further work will give quantitative methods of predicting ride severity from measures of the input vibrational conditions.

Now *you* try to isolate the principal matters, then try to write an abstract of the abstract. Check your results against the answers in appendix 8.3.1.

8.3.2

Select several key thoughts from among the principal matters. If you are writing an abstract of your own work, you've already done that. Your key thoughts are your group headings (sec. 2.1.423). If you're working on another's effort, you have to look closely; do not mistake a detail for a key item. Mrs. Pearl Parlett, then a bookkeeper in Cornell University's Entomology Department, did a fine job in noting the key thoughts, then reconstituting the pared-down but efficiently pointed abstract.

Original:

This report defines Department Use Codes, Allocations, Personnel Commitments, Department Allocation Summaries, and gives you information on which figures are to be inserted in the space provided.

Parlett:

This report defines Entomology Department accounting data for preparation of Monthly Summaries. It also gives sources for finding that information.

8.3.3

Using each keyword (or two) as a focus, confect sentences applicable to the paper's line of thought.

8.3.4

Put those new sentences into a logical sequence, so the sentences form a thought line simulating the original statement.

8.3.5

Add some small transitional words for seamless flow; this is adding quality.

8.3.6

Rewrite that newly formed statement once or twice more to consolidate and polish it. Try this twenty-four-word sentence (a list heading) plucked from an overinflated report. Try, in three rewrite steps, to cut it to its minimum. You may alter the form of any existing word, but may not add any new words. The process is in appendix 8.3.6. See how the correct answer evolved as *consider*.

> *It is the studied opinion of the writer that serious consideration should be given by the reader to the following items of informational material*:

8.4 SHORT FORM DIFFERS FROM SCOPE SUMMARY

Section 2.1.43 defines and details the purpose, content, and place of a Scope Summary. That is a target marker for where you hope to be at your project's completion. That is a memorandum to yourself, an aspiration of what you want your finished product to be. It is an undetailed end-result specification. A Scope Summary can be written only at the *start* of a project to give it direction.

An abstract (short form), as this chapter explains, is your finished document shrunk to scale via a prescribed process. Because it is extracted from the finished effort, it can be written only *after* the project is completed.

If your abstract reads reasonably close to what your Scope Summary aspired, you have written your document properly. If the two items coincide, you've done it perfectly. If they are somewhat divergent, you may be able to make some adjustments to the document. If they are different, you need to review the entire project to find where you went astray.

James R. Rundlett, CWO3, U.S. Coast Guard, filed an assigned memorandum for which he wrote an abstract. He checked that against the scope summary he wrote at the outset. He found a match; his memo was on target.

Abstract (after)

The need for an effective VHF High Level system remains. However, records required for effective management need extensive work.

Until now, equipment has been replaced as it failed, but we do not have a clear picture of future needs nor the quality of maintenance. A coordinated maintenance management plan will improve system reliability, provide scheduled equipment replacement, and reduce our dependence on crisis management.

Scope Summary (before)

This memo will *discuss* the management and modernization of the Ninth District VHF High Level system. Proposed changes will be supported by an *analysis* of costs, maintenance problems, and potential benefits.

8.5 EXECUTIVE SUMMARY

Over the past two decades, public demand for more and better goods and services from science, engineering, and industry has ballooned. Concomitant has been the sharp increase of voluminous and complex documentation, abetted by the consistent improvement of electronic word processing devices. That resultant growth of required reading combined with administrative demands has created time constraints for decision makers and directors. A fully detailed report for the record becomes too much; the abstract, too little. The workable middle ground is the Executive Summary.

Method: You must start with a completed and polished document, as detailed in chapter 2. Then use the method in section 8.3 to produce an abstract that you will use as both an overture to your document, and a trail guide for your Executive Summary.

Now, open your abstract and infuse as many pertinent details as you need to steer your reader toward the conclusion. That will evolve a reasonable miniature replica of the original work. Start with a clear, strong statement of your objective. Rewrite once or twice to consolidate.

A consumer products company assigned a group to study the economics of adding a new product; K. C. Baker wrote the twenty-three-page report. Next he wrote an abstract, then from them both he evolved an Executive Summary.

> Abstract
>
> This analysis develops the project economics, sensitizes risk levels, and discusses some concerns for the 30 oz. "G" Container tooling investment proposal. It suggests a logical and statistical analysis, and recommends a basis for evaluating the proposal.
>
> Executive Summary
>
> The 30 oz. "G" Container tooling proposal generates an internal rate of return in excess of 100%, and a payback of under two years on the incremental $75M investment.
>
> The returns are driven by an assumed 35% attained variable margin in year one, and 40% annual volume growth over the three-year introductory period.
>
> Failure to realize a minimum of 60% of targeted volumes *or* margins, however, will reduce project returns to an unacceptable level. Sales histories of existing companion products (e.g., 15 oz. and 25 oz. "G" Containers) show virtually flat performance over the past two years. This trend suggests poor performance for the proposed line.
>
> We recommend Marketing review volume and pricing assumptions thoroughly before forwarding the proposal for General Manager approval.

8.6 SUGGESTION

Using this chapter and some reasonable thought, you should be able to compose with confident comfort a creditable short form of whatever size your job demands and useful for your target receiver. Every professional can acquire and employ this necessary skill. To a fortunate few, excellence in *shrinking to scale* comes naturally. Some good examples from the outside literature come to mind: Lord Byron (George Noel Gordon) abruptly terminates his interminable sea of flotsam and jetsam with "Sunk, in short"; writer Alice McDermott encapsulates an insatiable grandma's ninety-year life into her short, short story, "Enough"; years ago in a waiting room's *Ladies Home Journal* copy, I read Inez M. Merrill's verse that skillfully condensed three generations into four short lines:

My newborn baby tumbles from my arms,
And toddles off to school.
A tall youth saunters home with her,
Their child brings me a shawl and stool.

Those examples have the unquantifiable characteristic defined in quiddity (sec. 8.2.4) as *quality* in writing.

BREVITY IS THE SOUL OF WIT.[1]

—Shakespeare

NOTE

1. *Hamlet*, act II, scene II, line 90, Polonius to Queen.

9
FORMATS

9.0

Some governmental agencies and many private firms and organizations have their own specific in-house written report formats. Also, commercial computer software is available that will structure a report for you. The key is adaptability. Is a given prestructured format sufficiently adaptable to your reason(s) for writing, to the subject material, and for the needs and wants of the target reader(s)?

For writers whose employers or retainers do not impose such constraints, or for those who operate independently, I offer a multipurpose generic format. This is *not* a one-size-fits-all setup. It requires tailoring to accommodate your specific project at hand. At a buffet table, you do not pile some of everything onto your plate; selectively you take what you want, and *leave* the rest. Do the same here. Choose from among the format structure headings that fit your project, and ignore the unsuitable ones. With few possible exceptions, no report includes them all.

9.1 A MULTIPURPOSE GENERIC REPORT STRUCTURE

Your report might evolve in a multipurpose generic report structure. The suggested entries are in ALL CAPITAL TYPE; *their explanations and working comments are in italics*. Use all or selected parts of this structure for your final draft *after* you have done all the drills in chapters 1 and 2. Then, using the skills acquired in chapters 3 through 7, *carefully* write your report employing as many prototype ENTRIES as you need to complete a document responsive to your assignment, A.

9.1.1 REPORT FORMAT PROTOTYPE

Front Matter

<u>TITLE:</u>
This should be a truncation of your subject statement.

CLASSIFICATION: *In government and some organizations this is mandatory. Those readers without "Secret" or higher classifications should read no further.*
DATE: *The date the report was issued.*
PRIORITY: *If requested by your organization.*
NUMBER OF ACTUAL REPORT PAGES:
NUMBER OF APPENDICES:
Important to tell the reader how many real pages and how many elucidatory pages.
DISPOSITION:
What your reader is to do with the paper after reading.

ADDRESS TO: *Primary target reader, source of assignment, with complete address.*
DISTRIBUTION: *List six or fewer. If more, start the appendix with it.*
FROM: *Originating organization, Chief, Author (Signatory), other names, aides, and help. Names for credit, and responsibility.*
SUBJECT: *Explicit expanded form of TITLE, from assignment.*
GIVEN: *Basic assignment information.*
KEY ANSWER: *"Yes," "No," "Perhaps," or whatever the conclusion is, in ten or fewer words.*
KEYWORDS: *Several subjective words for information retrieval. In many cases these may well be the major headings from your outline (sec. 2.1.423).*

<u>ABSTRACT:</u>
See Chapter 8.

<u>CONTENTS:</u>

Section No	Subtitle	Page Number
1.0	*Only major headings from outline.*	*Actual page numbers of finished, published document.*
2.0		
etc.		

GLOSSARY: *If you use more than six new words, list and define all. Fewer than six, define in text (in parentheses).*
BOUNDARY (or SPECIAL) CONDITIONS: *Depends upon the project.*

EQUIPMENT AND MATERIALS: *Depends upon the project.*
FORMULAS, ABBREVIATIONS, AND SYMBOLS: *Better here than in the text. Derivations in appendix.*

9.1.2 REPORT BODY

Be responsive to your assignment. Write the complete text following your outline (secs. 2.3.11, 3.0, and 4.0), adding depth and details from your research and acquired information (materials in research envelopes (sec. 2.2.6).

9.1.3 END MATTER

<u>RESULTS</u>: *If any, responsive to charge, fully explained and detailed.*
<u>CONCLUSIONS</u>: *Based on results, but only if required by assignment. If report is a group effort, and is not unanimous, the dissenting opinions must be included for the record.*
<u>RECOMMENDATIONS</u>: *Based on Conclusions, but only if required by the charge.*
<u>SIGNATURES</u>: *Personally by the writer. If required by the organization, also signed by a superior, for credit and responsibility.*
<u>APPENDIX</u>: *All pertinent material too detailed or cumbersome for the continuity of the text. Also interesting minutiae peripherally related to the subject.*
<u>ACKNOWLEDGMENT</u>: *All contributors, and anyone who was of any help. Names add verisimilitude, and build good will.*
<u>REFERENCES</u>: *Cite only those publications from which you have actually culled specific information, or quoted as verification of your statements. Include title, author, date, and page,* vide *typical entry here in* <u>Bibliography</u> *(see chap. 20).*
<u>BIBLIOGRAPHY</u>: *This list could be labeled alternatively* **ADDITIONAL READING**. *These are publications not necessarily seen or read by you but could be interesting to your reader, if he'd care to learn more. It should be posted in true bibliographic style (see chap. 20).*
<u>TABLES</u>: *And other substantive materials. Derivations too large for text.*
<u>ILLUSTRATIONS</u>: *All photographs, sketches, and graphics belong in the text body, as close to the referred item as production allows. Put them into the appendix only if local policy requires (albeit poor form). Key all graphics in the appendix to their referred items in the text (secs. 3.3.3 and 5.6).*

9.2 APPLICATION

Section 9.2.1 shows the front matter of a real report, written using selected elements from the format recommended in section 9.1. The complete report following that method is in appendix 9.2.1.

9.2.1 REPORT FORMAT PROTOTYPE

<u>**North Atlantic Gyre Recirculation**</u>

Date: Day, Month, Year
7 pages + 3 Appendix
Unclassified

<u>To</u>: Dr. Cheryl Peach
    Chief Scientist
    SSV *Corwith Cramer,* Cruise 141
    Sea Education Association

<u>From</u>: Christopher P.M.D. Morris
      Cornell University
      SSV *Corwith Cramer,* Cruise 141
      Sea Education Association

<u>Subject</u>: Geostrophic Flow in the North Atlantic: In Search of Gulf Stream Recirculation.

<u>Given:</u> The northeastern flow of the Gulf Stream water in the North Atlantic increases from its beginnings in the Straits of Florida almost threefold to where it pushes off the continent off the Carolinas. This additional water has to come from somewhere.

<u>Key Answer</u>: The additional water comes from recirculation within the North Atlantic Gyre itself.

<u>Key Terms</u>: Geostrophic flow, gyre, Coriolis effect, Sverdrup, recirculation transport, hydrowire, rosette, hydrocast, specific volume anomaly, integrated flow, vertical pressure gradient, Ekman transport, geopotential anomaly, perpendicular transport.

<u>Abstract</u>: This project was designed to seek geostrophic flow recirculation from the North Atlantic Gyre system to the Gulf Stream. Depth, temperature, salinity, and density data were recorded by CTD cast at selected stations along the cruise track of the SSV *Corwith Crame*r Cruise 141. We processed data to find incremental geostrophic flow and total flow for stations covering the area between casts. The study detected the presence of the Gulf Stream, subtropical convergence zone, and geostrophic recirculation.

CONTENTS

Section	Subtopic	Page
1.0	Background	1
2.0	Methodology	3
3.0	Results	5
4.0	Discussion	6
5.0	Conclusions	6
6.0	Works Cited	7
7.0	Acknowledgments	7
A	Appendix	8

(REPORT BODY: see app. 9.2.1)

<u>Abbreviations</u>: SSV—Sailing School Vessel
CTD—Conductivity, Temperature, and Density sensor
Sv—Sverdrup (amount of water transport equal to 1.0×10^6 m³ per second)

<u>Equipment</u>: see section 2.2
<u>Conclusion</u>: see section 5.0

SIGNED
Christopher P. M. D. Morris
Cornell University
Sea Education Association
Corwith Cramer Cruise 141

9.3 LETTER REPORTS

A rare few practitioners in today's hasty, computerized world will take even the shortest time to smell the flowers, or read the most competently composed complete report. For those *e-mailaholics* who insist on instant mental digestion, you *can* gratify their executive or client cravings with a Letter Report.

All you need is a modicum of perspicacity, along with an ability to combine the instructions on reports and short form (chap. 8), to produce a good Letter Report. You can regard it as some folks do a bat: a something between a bird and a rat. A letter report is a document between a full report (chaps. 2 and 3) and an Executive Summary (sec. 8.5).

Specifically, first construct your report, because you will need that to complete your assignment, and for the record. Next, use the technique for writing an Executive Summary (sec. 8.5). Then enlarge the Executive Summary (with factual material, not bloat) to where it occupies two pages, no more, no less. That product is your Letter Report.

Lane Holstun at West Point Pepperell finished an assigned, in-depth investigation of a safety matter required by the U.S. Environmental Protection Agency (EPA). He filed his finished document as assigned, then had to submit a letter report to keep management apprised of his activity. In reading his two-page letter report (sec. 9.3.1), you will find all the undetailed information the target reader would need at that point. This example is so well written that in reading the two-page text, in your mind's eye you can see clearly where the outline index numbers would be if they were to have been included.

9.3.1 LETTER REPORT PROTOTYPE

WEST POINT PEPPERELL
Lanett, Alabama

Date
Page 1 of 2
Three-Year Revised Oil Spill
Prevention, Countermeasure Plan
for Granton Plant

From: Lane Holstun
To: Mr. John Smith

Abstract:

This SPCC plan was prepared by Corporate and Facility engineers at the direction of Management having authority to commit funds to implement the plan. Potential spill areas, countermeasures, inspections, oil handling, and facility security are defined to explain compliance with EPA requirements.

POTENTIAL SPILLS:

Granton Plant has three areas where we keep oil.

55 gallon drums of lubricating oil are in the shop. On the northeastern end of the mill yard is a 10,000 gallon aboveground tank storing No. 2 fuel oil. Substations have six (6) oil-filled transformers containing 20 gallons each. Potential spills would work toward a stream bordering the northernmost side of the plant. The shop floor would hold any oil spill there in its entirety. The aboveground 10,000 gallon tank contents would have to flow 150 yards northerly on the ground to reach the stream, if the four foot wall around the tank were not there. Granton Plant has no mobile storage tanks. There is a two inch pipeline from the 10,000 gallon tank underground approximately 100 feet to the boiler. This pipeline is not traveled over by any vehicles.

MEANS OF CONTAINMENT:

There are usually two or three 55 gallon drums of lubricating oil in the shop. The drums are placed in conspicuous places that have no chance of reaching a drain. Oil would be removed by absorbent material. A four foot wall of 8" blocks has been constructed around the 10,000 gallon tank and has a holding capacity of 10,700 gallons. Oil removal from within this wall would be done by connecting a pump and a hose to a one and one-half inch hand-operated gate valve at a low accessible point. Water treated at this plant is regularly monitored by a West Point Pepperell testing lab. Samples are collected at a constructed weir situated in the stream below all plant entries. Results of postings are recorded and sent to the State of Georgia. The six (6) 50 KVA transformers are where spills will not reach the stream or drain. Absorbent material is on hand to collect spills.

INSPECTIONS:

The storage areas, drums, tanks, containment devices, valves, pipelines, and security are inspected monthly by the Plant Maintenance Engineer. Any discrepancy is recorded during the inspection, and repair is done immediately, if there are discrepancies. Tankers bringing oil in are inspected and records of inspections by the Engineer are kept on file.

HANDLING SAFETY:

Trucks to pump oil into tank have trailer wheels chocked before connecting hose to fill line to prevent a premature disconnection by truck rolling off. Our supervisors and employees have been instructed to guard against any possible cause of spillage. Conspicuous places, bulletin boards, have a bulletin addressing employees with directions telling whom to call, phone numbers, what to do, and what information to supply when reporting a spill. We have a qualified maintenance man standing by to assist while tankers are unloaded. This man will, by hand signal and voice, have contact with truck pump operator to prevent overfilling tank.

SECURITY:

The entire plant is surrounded by an eight foot chain-link fence. Security is maintained by guards at locked fence gates and the boiler room where the locked oil pumps are. The drums in the shop are locked within the plant itself. The mill yard is lighted, the guards make rounds, and constantly inspect for leaks with orders of whom to call in the event of finding a leak. Guards are under the direction of the Plant Engineer who is responsible to the Mill Manager for implementation of this plan.

SIGNED
Lane Holstun

LH/eg
  cc: AGB
       BCW
       CCB

9.4 PARALLEL STRUCTURE

Seldom do you find a viable writing procedure available that is usable, writer friendly, reader friendly, and logical. Parallel structure is that kind of a report format, wherein everyone wins. Easy to write and easy to read, it is most applicable in large project reporting, but it does not work for all projects; look at the whole picture first.

Consider this one as an ideal prototypical situation as it actually developed. Before its new building construction, Cornell University Bio-Sciences planned to rework and modernize its aging laboratory. As presented to administration, the extensive report plan began with an overall "as is" view of the archaic facility existing in the Stimson Hall attic. Next came all the proposed changes with a time line. The report concluded with the financial aspects (sec. 9.4.1), adding an amply detailed appendix to keep the required minutiae from log-jamming the narrative.

To make all that material easier to organize, interrelate, and present understandably, the writer, Robert Geyer, put the same specific entry into the same position in each report section. For instance, regard his report section 3.1, Initial Costs. Focus on item *3.1.62 Steam Lines - $7,000*. All there is to read is that one line. Should you want to know what is now on site, you must read his report section 1.1.62 for a detailed description of existing steam lines. If you want to know what Geyer's group envisions as replacement, read his report section 2.1.62, Proposed Changes. To ascertain what those improvements will cost, see section 3.1.62, Financing, where the only information presented is the bottom line, $7,000. $7,000?! Finally, to learn the minutely detailed breakdown (to the last washer on a bolt), it is all in *his* appendix 3.1.62. It is all in there because it is too voluminous to place in the text without interrupting the narrative flow. In this format, a reader is able to isolate and read with ease and comfort (sec. 9.3.1) all he'd want to know about a specific project item by seeking any section (see sec. 1.6.2) without having to go *cherry picking* throughout the full report.

9.4.1 PROTOTYPE OF PARALLEL REPORT STRUCTURE

3.0 Finance

    3.1 Initial Costs $20,000

Initial renovation costs will total $20,000 based on a project start this June. Funds for this renovation will not become available until this July when the new budget for the year is approved. The construction of the Behrman Biology Center has exhausted all funds for new projects this year, and the Deferred Maintenance Program has several pressing and expensive repairs scheduled that must take precedence over this project. That will drain the current budget of funds for any other renovations, thus the next is the earliest practical starting time for this project. The funds for the renovations considered in this report should be drawn from the Endowed budget and the Building Fund for maintenance on a 50%/50% basis. I have broken down the total cost into what each facility will cost to renovate as outlined in section 2.0. All finance details are in appendix 3.1.62.

    3.1.1 Photoperiod Animal Facility $5,000
        3.1.11 Room structure $3,800
        3.1.12 Air conditioning $800
        3.1.13 Electrical $200
        3.1.14 Moving $200
    3.1.2 Autoclave $0
    3.1.3 Dishwashing Facility $0
    3.1.4 Vertebrate Zoology Storage $4,500
        3.1.41 Room structure $4,000
        3.1.42 Moving $500
    3.1.5 General Storage—Moving $500
    3.1.6 Distilled Water Facility $7,500
        3.1.61 Removal and reinstallation $400
        3.1.62 Steam lines $7,000
        3.1.63 Moving $100
    3.1.7 Distilled Water Taps $2,500
        3.1.71 Pipelines $2,200
        3.1.72 Taps $300
    3.1.8 Machine Shop $0

Robert Geyer, Mgr. Lab. Service
Cornell University Bio-Sciences

9.4.2 TRACKING STEAM LINE INFORMATION

The key to finding desired information on steam lines in that report would be to follow this route:

1.0 **EXISTING CONDITIONS**
    1.1.62 Steam lines (full current in situ detail description)

2.0 **PROPOSED CHANGES**
    2.1.62 Steam lines (changes described in detail)

3.0 **FINANCE**
3.1.62 Steam lines (bottom line only) $7,000

APPENDIX
3.1.62 Steam lines (complete, detailed proposal breakdown)

9.5

None of the formats illustrated in this chapter is an inviolate cookie-cutter template. They are here for you to think about them; think about their use for writer ease, and consequently for reader understanding and comfort, in communicating professional ideas.

FORM EVER FOLLOWS FUNCTION.

—Louis Henri Sullivan[1]

NOTE

1. "The Tall Office Building . . . " *Lippincott's Magazine*, March 1896.

10
EXPANDING OUTLINE AND TEXT

10.0 ENHANCEMENT WITHOUT EMBELLISHMENT

While you were working your way through the first nine chapters, hopefully you completed one or two useful documents at the same time and you feel at ease with the procedure. You have arrived at a *plateau*, halfway to being able to issue professional-grade communications. After this chapter, you will be able to shape your work to meet the exact needs and wants of different readers and different occasions by enlarging or diminishing presentations of the *same* basic materials.

From chapter 8 you learned the route to *shorten* a document while preserving its original senses of substance and quality. Here you will acquire the skill to *enrich* your document at appropriate selected places. *Then*, in chapter 11, you'll learn where, when, and how to expand or reduce your text to comply with the audience's requirements.

10.1 CAUTIONS

Some *wrong* ways to add more content are to inflate the text with deadhead expressions or quote unnecessary references. (Both are contrary to this book's basic premises.) Do not inject large bodies of minutiae into the existing text; that is stifling to any reader, and sidetracks the flow of your theme. Adding a large parcel at the end may appear to simplify the chore, but it becomes a loose caboose to your train of thought, no matter how precise your cross-indexing may be.

If conditions are such that a large parcel of detailed "boilerplate" material is demanded for a pinpointed secondary reader, then reference note it in the appropriate text place and put the packet into the appendix. There, those who have a need or interest may delve into it; thus, your document is not lacking, and that added (and stored) material does not disrupt your narrative flow.

10.2 MECHANICS

The best way to spread and enrich your text and still retain its consistency is to *expand your outline* by infusing needed new notions, then fill in with meaningful details.

First try: Go back to your original query sheets and face sheets to find any unanswered questions or discarded answers (sec. 2.3.1). Generate and research more questions to acquire more in-depth, hard news. Carefully create fitting new inert umbrella headings (fig. 10.2). Now integrate the new items into your original outline (see fig. 2.4) where appropriate; your expanded outline should have a similar structure to Figure 10.2. In the figure, the original is shown in normal font, while the newly added items and six new umbrella headings appear in *italics*. Now rewrite the project's section 5 from the expanded outline, infusing the new material, and you have your expanded section. You are now prepared with the ability to handle the rest of this text.

FIGURE 10.2. EXPANDED OUTLINE

```
5.0 Marketing
    5.1 Eggs
        5.1.1 Trade Areas
            5.1.11 Processors
            5.1.12 Wholesale
        5.1.2 Market Areas
            5.1.21 Chain
            5.1.22 Jobbers
            5.1.23 Individual
    5.2 Chickens
        5.2.1 Direct Sales
            5.2.11 Private
            5.2.12 Public
        5.2.2 Excl. Sales
            5.2.21 Contract
            5.2.22 Transient
    5.3 Feathers
        5.3.1 Industrial
            5.3.11 Bedding
            5.3.12 Mattresses
        5.3.2 Commercial
            5.3.21 Decoration
            5.3.22 Hats
    5.4 Litter
        5.4.1 Chemical Reduction
            5.4.11 Fertilizer
        5.4.2 Sales
            5.4.21 Private
            5.4.22 Government
```

ADDED MATERIAL MUST GIVE THE READER MORE <u>USEABLE</u> INFORMATION. EVERY ADDED WORD MUST <u>WORK</u>.

11
FOCUS ANALYSIS

11.0 VOLUME VERSUS VALUE

By this time *your* own document is finished and refined. You should be able to recognize the existence of any imbalance in major section lengths. That probably was caused by more available researched material for what may be a section of lesser importance, vis-à-vis fewer source items for another principal heading perhaps more vital to the *whole* project. Now is the time for recension. But you must expand or reduce the affected sections according to a prescribed method, not by reckoning or whim. Focus analysis is the reliable way to determine *where* and *how much* to cut or fill.

Focus analysis is accurate, though not precise. Unlike medicine, engineering, or science, technical writing, although rooted in truth, is subject to human frailties. Ostensibly we are all human beings, thus an *accurate* although *imprecise* method is the best possible attainment for this objective.

11.1 THE CHART

You have followed from scratch the prototype Chicken Farm project. Now take it a big step further; proportion it. Draw a six-column chart layout like Figure 11.1, then start to fill it in.

11.1.1 COLUMN 1

Easy. List the major group headings in order from your outline.

11.1.2 THINK!

In building that outline (sec. 2.4.2) you selected that heading arrangement for your writing ease, not for order of importance to the project. Now you assign each heading a priority number, an order of importance. (The only exception is if you had chosen the order of importance arrangement at the outset, then column 1 and column 2 will be identical.) What you can sell is prime to the whole (1); it must exceed the combined costs of Plant (2), and Operating Expense (2); the Chickens (3) and Plant (3) are cofacilitators.

11.1.3

Your complete report (100%) is 153 pages. Thus, in column 3, enter the *actual* number of pages each section contains, being sure the column total equals 153. Now figure the page percentage each section contains, being sure the percentage total equals 100% (e.g., 3.0 Chickens has 61 pages, and 61 is 40% of 153).

11.1.4

For the Chicken Farm project, because of its 153-page length, the most workable unit for comparison measure is *page* count. For short papers of a few pages, a *word* count is best. For longer papers, a *line* count is manageable. Use *pages* for papers of fifty pages or more. It depends on what is a

FIGURE 11.1. FOCUS ANALYSIS CHART LAYOUT

MAJOR GROUP HEADING (FROM OUTLINE) (11.1.1)	ORDER OF VALUE (PRIORITY, IMPORTANCE) (11.1.2)	PRORATE (%) OF TOTAL PAGES VS. ACTUAL # OF PAGES 153		TO DO (11.1.5)	COMMENT (11.1.6)
		ACTUAL # ACTUAL (%) (11.1.3)	DESIRED # DESIRED (%) (11.1.4)		
1.0 Funds	2	30 (20%)	30 (20%)	0	Leave as is.
2.0 Plant	3	15 (10%)	15 (10%)	0	Leave as is.
3.0 Chickens	3	61 (40%)	15 (10%)	Vast cut to fit.	Use short form technique (sec. 8.4).
4.0 Operations	2	24 (15%)	31 (20%)	Add 5% (7 pages), leave as is at first.	Adjust later if warranted.
5.0 Marketing	1	23 (15%)	62 (40%)	Increase by 38.	Increase via expanded O/L and rewrite (sec. 10.0).

workable *scale*. I've known writers who counted paragraphs. A priori, that is vastly inaccurate because every paragraph is of a different length, whereas words, lines, and pages are reasonably constant, thus comparable.

In column 4 you work the other way, guided by the *order of importance* established in column 2. You *assign* the appropriate percentage of the whole you want in each section. This case was easy: first priority got 40%; the two second-priority sections each got half that number, 20%; the two remaining third-priority sections got 10% each. Column 4 equates the percentage allotment to the number of pages it ought to contain. Now you can see the section length imbalances with regard to their value to the whole. Be sure your adjusted percentages total 100 and page numbers total ±153.

11.1.5

Columns 5 and 6: Recognize what you must do with existing numbers of pages in column 3 to make them coincide with the *desired* numbers in column 4. Do nothing with sections 1.0 and 2.0. You need a vast cut in section 3.0, but not by lopping off pages. Use the short-form process of chapter 8.

For 4.0 Operations, calculations suggest adding 7 pages (5%). Leave it alone for now; this is your cushion. Add more later *if* you need, after sections 3.0 and 5.0 are adjusted. Section 5.0 requires an increase of 38 pages (25%). You handle that by expanding the outline (chapter 10), then rewriting the section.

11.1.6

All that is focus analysis, and how to apply it. It may look like a lot of work on what might have seemed a finished document, but it makes the difference between a good report and an excellent report.

GOVERN THE VOLUME OF YOUR COMMUNICATION BY THE
VALUE OF THE MESSAGE IT CARRIES.

12

ALTERNATIVE SPECIALIZED FORMATS

12.0 ENHANCEMENT

This is an optional chapter, not meant for all writers. You may skip it without appreciable loss. It deals with a sophistication requiring common sense, perspicacity, good judgment, skill, and much practice. But it is worth trying.

On first reading, this chapter may appear as if it contradicts the earliest lessons of starting from nothing more than your reason for writing and your real audience (sec. 2.1). Then, from your open-ended questions, develop your principal matters. You do *not* preimpose outline thoughts to fill in around them, because you won't be able to know what you might have missed. This current chapter covers the opposite approach, except now one major governing condition has *changed*.

At this stage you will have gone through *all* the protocol of chapters 2, 3, 4, 9, 10, and 11, and now have a completely focused, finished document that should satisfy all your levels of readers. Your project can and will suffice as is. For those among you who demand a product a cut above *finished*, I offer a "cheese on the spaghetti" advancement for enhancement.

Since you are *not* starting from ground zero, and have a polished product, you can rearrange the content presentation just as you added inert umbrella headings during development.

12.1 APPLICATIONS

Figure 12.1 depicts schematically the techno-professional areas where you write to, from, and within. Simply stated: Scientists are highly educated probers into the universe who seek consistent answers to physical world happenings; engineers are the realists who use technically proven ideas to build the tangible world around us; practitioners and paraprofessionals are trained in pragmatic technology to aid the first two groups; and technicians put all the parts together and make it work.

12.2 ONWARD AND UPWARD

If your document emerges triumphant through all the preceding chapters, and it achieves the result you sought, to answer your assignment, it is *done*. Let it be. Should you decide to embellish it to a higher polish, you might try reshaping the existing content (no further adds, no more deletions) along the lines of Figure 12.2 for the higher reader groups, or Figure 12.3 for all others, or, using your judgment and experience, any reasonable median variation to suit the specific situation at hand.

Your first, even third, try might not yield sterling success with this technique. It demands all those conditions listed in section 12.0, plus a lot of practice. When you get there you'll know it, and be glad you did.

FIGURE 12.1. PROFESSIONAL INTEREST AREAS

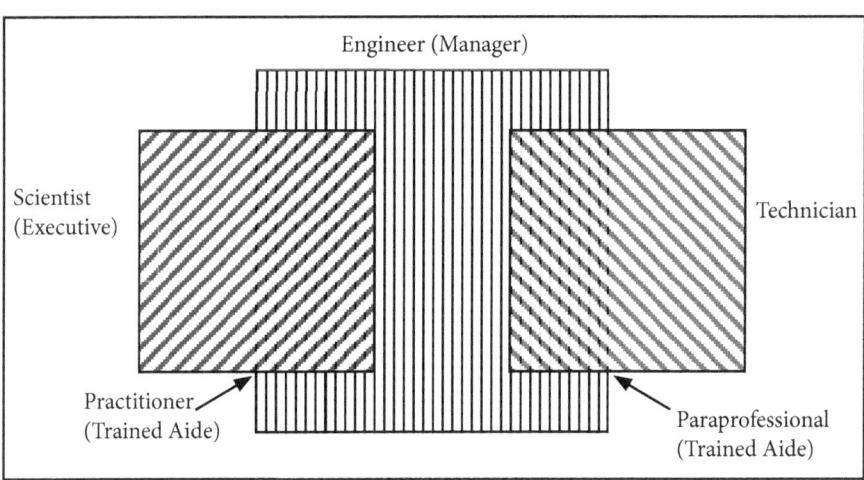

ALTERNATIVE SPECIALIZED FORMATS 113

FIGURE 12.2. GUIDELINES FOR HIGHER LEVEL READERS

> 1.0 Hypothesis
> 2.0 Evolved Theory
> 3.0 Experimental Trial
>   3.1 Consistent Repetitions
> 4.0 Evaluations
> 5.0 Recommendations
>
> Appendix

FIGURE 12.3. GUIDELINES FOR GENERAL READERS

> 1.0 End Result Requirements
> 2.0 History
> 3.0 Design
>   3.1 Mockup
> 3.2 Prototype
> 4.0 Field Trials
>   4.1 Consistent Results
> 5.0 Evaluations
>   5.1 Changes
> 6.0 Conclusion
>
> Appendix

TO EXCEL, A DOCUMENT MUST ANSWER THE NEEDS AND WANTS OF ITS READER, WHILE GRATIFYING THE PROFESSIONAL EXPRESSION OF ITS WRITER.

13
ORAL PRESENTATIONS

13.0

Your oral presentation lives or dies on the strength of your preparation. Your *delivery* is just the nozzle; your *material* is the water; but your *preparation*, the hose, is what carries it all through smoothly from start to finish. Three considerations are essential to planning your oral presentation:

13.0.1

Scope; the breadth of your subject coverage.

13.0.2

Depth; to which your broad coverage delves.

13.0.3

Time; how long you should talk to accomplish the first two adequately.

13.1

Your first thought in beginning to prepare any communication (including oral presentations) is *why write*? Any of the six reasons detailed in sections 1.4 and 2.4, along with the expected audience, *to whom*, determine the *what*, the content of your oral presentation.

13.1.1

The presentation could be in person with response to a small audience; to your boss about a raise or a promotion; to peers, your boss, or management about a new idea; or to a client, promoting a product.

Or you can accomplish all those via teleconferencing or video-conferencing. Those are all instances of when *the audience determines the subject*.

13.1.2

The presentation could be in person (only), with or without response. It could be to a large audience as a lecture on a specific topic; to stockholders regarding a management decision; a political speech (pro or con) about a candidate or an issue; to a community about a change or development; or a sermon. Those are all instances of when *the subject determines the audience*.

13.1.3

For any occasion, your oral presentation should include facts, ideas, and opinions. But to justify their listening, it all must be of *value* to your audience.

13.2 PREPARATION OF SELF

When you stand before an audience of any size, in effect you are "naked in Macy's window on a Saturday afternoon during the Christmas rush."[1] To be of any value to that audience, and to yourself, you have to be on top of your game. Believe in what you say, and in yourself. Then you will convince your listeners.

13.2.1

Start by standing about five feet in front of a full-length mirror, then say and gesticulate something funny and something profound. What you see is what your audience gets. Evaluate that picture honestly. Take caution from the two Wills. Shakespeare wrote, "This above all: to thine own self be true"[2] Will Rogers[3] more succinctly quipped, "Don't kid yourself."

13.2.2

Make serious notes of what you saw in the mirror. Do a few retakes. Are you looking around at the entire assemblage? Do you move and gesture freely? How is your enunciation? Your inflection? Hone them all, and accentuate all your positive attributes. Spot the flaws: no looking down, no fiddling with the necktie or anything on the lectern, don't pull on your ear, and cut the "um . . . uh . . ."

13.2.3

Your appearance should be impeccable: suit pressed, shirt ironed, shoes shined, and no loud neckties. Dress appropriately for the audience, but *don't* presume to be one of them *if* you are not; that can be resented. When you talk to stevedores, wear casual clothes, but *do not* dress like a stevedore.

Feel at ease on the podium. *Sincerity* is the most positive impression you can make.

13.3 PREPARATION OF CONTENT

Making an ordinary oral presentation into a superb oral presentation takes long and careful planning and preparation. I've learned over the past half century of the genuine need for a finished technical paper as a proper, stable foundation for your oral presentation. Organize your project following the trail posted and blazed in *all* this volume's foregoing chapters. Then polish and produce it for an audience takeaway even though it may be published in a journal or proceedings. Make paper copies available to all as they leave, *after* your presentation. Handing it out before allows them to read and diverts their attention from your oral presentation.

13.3.1

Condense your detailed, expanded outline into a keyword listing in large type, triple spaced, easy for you to check to stay on discourse course.

13.3.2

Select dynamic graphics to fit available facilities. Many speakers have their materials on laptop computers. An odd few effectively still use flip pads on easels or dry erase markers on a whiteboard. Those latter two can be ideal for talks involving a lot of math, other equations, or line drawings.

13.3.3

PowerPoint® has almost become the standard for oral presentations. It is colorful, dynamic, and can include appropriate background music. But PowerPoint carries some cautions. If the PowerPoint production is too entertaining, your audience may lose the message in the dynamics. Also, it may diminish the affect and effect of *you* in the presentation. Use this good tool cautiously and judiciously.

13.4 PREPARATION FOR DELIVERY

13.4.1

Reread your finished paper to *learn* but *not* memorize the now-familiar content.

13.4.2

Prepare a 5" × 7" prompter card and print vertically, in block letters, keywords from the text to jog your mind for what's next.

13.4.3

Some hours before your balloon goes up, inspect the oral presentation site to "case the joint." Check the seating. Look left and right. Can you see every seat? Check the rostrum (podium or lectern) for best position and height for yourself. Will you have to stoop, or will you be lost from the audience's view?

13.4.4

Check the working order of your visual aids: Is your computer operating well? Can you access and run your PowerPoint file (sec. 13.3.3)? Is the slide projector functioning and are the slides in the proper sequence?

13.4.5

Check the podium microphone, or the lapel mic and cord. If all those things are satisfactory, *you* are ready to go.

13.5 THE TWO- TO FIVE-MINUTE "QUICKIE"

You are seated at the head dinner table and are asked to respond to your introduction; you want to voice your opinion at a town meeting; you want to contribute more information not included in another's oral presentation (but *never* just to vent). Most of those occasions provide two or three minutes for "impromptu" statements, and you *know* beforehand you want to speak: be prepared. More than three minutes is in bad taste; less than two allows no time for a cogent statement. Optimum is about three minutes.

13.5.1

Experience has evolved this rule of thumb procedure: Make a list of a dozen or so items you believe are paramount to the subject under discussion; pare that list to the six items most important to the audience; reduce that six to one keyword each; list those six words in a column on the blank back of a normal business card to hold in your palm. Then talk freely for 30 seconds about each one of your six prompts, say "Thank you," and sit down.

13.5.2 A FIRST VISIT SALES CALL

The technique for hurdling entrance barriers was detailed in section 2.1.44. Once at the actual interview, use a prepared short form (sec. 13.5.1) statement of your case on the back of your business card, for him to keep. Then let him conduct his own question-and-answer period. Be prepared!

13.6 INTRODUCING A SPEAKER

You *must* realize that it is *his* time in the sun, *not* yours; act accordingly. It is easy if he is someone you know, thus know about. If you are not acquainted, you should speak to him before his oral presentation to learn his feelings about the landmarks of his career. You might have gotten it all from another source, but *not* every speaker wants it all aired out.

13.6.1

Don't drone on at length, you will be eating into *his* time. And *never* read an introduction from a paper; it is deadly. Use the back of a business card to list keywords, then tell the high points *he* prefers, then sit down.

13.6.2

Depending on the occasion and the assemblage, I might have quipped, "As Mark Antony said to Cleopatra, 'I didn't come here to talk'" (app. 13.6.2).

13.7 DELIVERY DON'TS

13.7.1

Don't read your paper. They won't stay, because they can read it for themselves later, and do something useful now (sec. 13.8.1).

13.7.2

Don't use too many slides. That keeps the room dark too long. No slides with too much writing, nor too small lettering. It becomes frustrating.

13.7.3

Don't write whole sentences on boards. Takes time from the talk. They came to listen, not to read.

13.7.4

Don't drag along, then speed to finish. Practice a uniform, listenable pace without drawl or speed (secs. 13.8.2 and 13.8.7).

13.7.5

Don't bark into the microphone. Keep about a full handspan's distance between your mouth and a stationary microphone. With a lapel mic, talk normally.

13.7.6

Don't clown or tell too many jokes. They'll remember the quips and not the talk. You do need a few well-salted short jokes to keep it alive (app. 13.7.6).

13.7.7

Don't turn your back to the audience: they'll walk out. Maintain eye contact with different listeners throughout.

13.7.8

Don't use all your material. Save some for detailed answers in the question period. Remember to leave time for questions.

13.7.9

Don't act; be sincere. Reread the last line of section 13.2.3.

13.8 DELIVERY DOS

13.8.1

Tell what your paper is about. Tell what they can learn from reading the paper *later*; for now, tell them how you did it, with reference to the paper.

13.8.2

Relax and talk naturally, as if you were one-on-one over good coffee[4] (sec. 13.7.4).

13.8.3

Use dynamic demonstration devices selectively. A speaker once talked about quick-setting chemical grouting. He produced a 4" diameter closed jar of clear liquid, shaken for veracity. Removing the lid, he put the open jar on the podium proclaiming it would set up in six minutes, then went on to finish his discourse. Six minutes later he picked up the jar and inverted it. The clear liquid was solid and stuck to the jar bottom. He said "Q.E.D.,[5] thank you," and departed.

13.8.4

Keep the room, talk, and atmosphere light. Enough said.

13.8.5

Pause frequently to let your ideas be absorbed. Give them a little time to think.

13.8.6

Make your whole thought clear to all. Offer an idea, then tell them the benefits (and risks, if any). Tell them why and how you did something and what it means. Give instances.

13.8.7

Pace yourself to get the whole thing into the allotted time (sec. 13.7.4).

13.8.8

Save some material for the question period (sec. 13.7.8).

13.8.9

Conclude with a short summary and restatement of your premise, then depart gracefully. Don't trip while leaving the lectern or drop your papers.

13.9 GOOD ORAL PRESENTATIONS

Jennifer Weill, the Director of Education for the Nanobiotechnology Center at Cornell University, spoke to an Ithaca Cornell Alumni meeting about "Nanobiotech and its meaning to today's society." She prepared a paper for the oral presentation, but did not take it to the lectern. She proceeded to talk clearly, precisely, and freely for a full hour without ever referring to a note; that is how well versed Weill was in her subject. The intelligent, nonscientific audience left impressed, with an encouraging view about the subject's future.

New York's new governor, David Paterson, addressed a National Press Club broadcast luncheon with a lucid, informative forty-five-minute presentation with *neither* prompting nor notes scribbled "on the cuff."

Many libraries offer various collections of the "World's Greatest Orations." I believe the all-time *greatest*, for length, documented fact, emotion, and *sincerity* is President Abraham Lincoln's *Gettysburg Address*.[6] Read it again, I'm sure you will agree.

> THE AIM OF FORENSIC ORATORY IS TO TEACH,
> TO DELIGHT, TO MOVE.
>
> —CICERO[7]

NOTES

1. The originator of this quotation is unknown.
2. *Hamlet*, Act 1, Scene III, Line 78, Polonius to Laertes.
3. Will(iam Penn Adair) Rogers (1879–1935), *Cowboy Philosopher*.
4. Or tea.
5. Q.E.D. is the abbreviation for *quo erat demonstrandum*, or that which was to be proved, was.
6. Abraham Lincoln's *Gettysburg Address* was delivered on November 19, 1863.
7. Marcus Tullius Cicero, 52 BC. *De Optimo Genere Oratorum*.

14

AUTHORSHIP

14.0

Authorship is like sculpture. Both are comparable forms of art, for the *talented,* and for the *skilled.* Both demand self-esteem and pride.

14.1

If you are a *talented* sculptor, you look at a block of marble and "see" your finished statue embedded in the block. Your job then is to chip away at the engulfing marble until you get down to your envisioned statue. Once you've liberated it from the excess material, you polish it and it becomes exhibitable.

As a *talented* author, you have your finished article in your mind's eye. Then you handwrite or tap it out on your computer as you "see" it, tossing out excess verbiage. Then you polish off the "flyspecks," and your article becomes publishable.

14.2

If you are a *skilled* sculptor, you sketch your envisioned statue. From that you build a full-sized, stiff-wire skeleton frame and beef it out with clay. *Skillfully* you sculpt and refine your statue from the rough clay. Around that you make a mold and fill it with your fluid material. When it sets, remove your casting, polish it, and your statue becomes exhibitable.

As a *skilled* author with your finished article in mind, you follow all the steps in the now-familiar route advocated in the preceding twelve chapters herein. That process should yield a substantial, serviceable report or paper (sec. 14.3). *Skillfully* then, beyond that, you reshape that product into a publishable article.

14.3

The process prior to your skilled reshaping into an article may be the extent a capable writer of reports and papers can attain. Perfecting and fine-tuning those skills will make an ambitious good professional into a better expressed professional, worth more per annum to your employer, to your client, and to yourself.

Many practitioners agree that constant rehoning of the directions of the first dozen chapters here will give them sufficient professional expression. They go on to perfect additional expertise in other channels.

14.4

In the preface I quoted my sagacious father as having taught me to "Profit from someone else's experience. If you can't, then profit from your own." Toward helping you do that, I share a few of my more remembered learning experiences as a young writer trying to succeed in this highly competitive field.

INSTANCE:

On a European tour with Overseas Press Club colleagues, I was bussed from West to East Berlin airports by the GDR. The bus paused at a construction job where the guide hawked their building efficiency and noted the brick masons using a newly devised method instead of the conventional Flemish bond. I was forbidden to make any photographs, but was permitted to make a sketch. Later, I sold that drawing along with a brief comment to a construction magazine for $15. Jack Harrison Pollock, an author of merit, also wrote a piece about that project. It became the cover article for an issue of *Parade* magazine and won considerable recognition. What I saw was a construction detail. What he saw was the building of the Berlin Wall; perspicacity.

LESSON:

Don't get caught up in a detail: be perspicacious, look around and see the whole picture, and its consequences.

14.5

INSTANCE:

A major construction magazine offered me $800 to write an eight-installment series on earth compaction, to run January through August of the next year. Eagerly I accepted the assignment with a handshake.

Monthly, in the prescribed lead time, my pieces were accepted and paid. With minimal editing they were run in the magazine on entire, consecutive full pages without advertising interruption. I was proud of my work. Six months later, eliminating two chapters, the other six were combined into a slick pamphlet using the same full-page plates. Over its lifetime that pamphlet sold 80,000 copies at fifty cents each. (One was pointed out to me in the library at the civil engineering school of the University of Moscow.) When I pressed the publisher for my royalty share, I was told I had sold to the magazine the eight articles for which I was paid. They were the copyright owner of the magazine content, and it was their property to do with as they chose. I had been paid for what I did.

LESSON:

When you take a big assignment, article, or book, look down the road. Discuss with your assignor who owns the rights to your intellectual property. Ask about first rights, and long-term implications. Publishers as a rule are fair-minded, ethical people, who for the most part deal in good faith. But not everyone is perfect.

14.6

INSTANCE:

My older son, now a successful independent writer, came in second in a New York State essay contest among high school students on the Franklin Delano Roosevelt centennial, about the FDR legacy. Whereas all other entries presumably presented those events in chronological order, Gregory borrowed an idea from my Cornell classmate Kurt Vonnegut's best-selling novel *Slaughterhouse Five*. Instead of unfolding his story in chronological order, Vonnegut leaped ahead in time for some events, then fell back for others as it suited the continuity. Gregory wrote his "E Chaos Ordo" as what we had when FDR took office and what he did, 1933; what brought it all about, 1929; what we have now, 1983; and FDR's legacy, from 1945—an arrangement that truly spotlighted the subject. I was told he really won, but the judges opined that he used too many "big words" and ironies, and may have had help. Obviously the judges hadn't read many of FDR's writings (sec. 18.5 and app. 18.5), nor were they fans of Cole Porter.

LESSON:

Chronological order is not always the best way of telling history (sec. 16.7.7); logical order may be better. Write to the estimated intellectual level of your audience.

14.7

INSTANCE:

As a native New Yorker, I was put onto an American Society of Civil Engineers (ASCE) committee whose members would show foreign visiting engineers around the engineered construction wonders of New York. After I had led many tours, the notion tumbled to me that plain Americans from Hawaii, Alaska, everywhere, might be fascinated with those useful accomplishments also. I proposed an article for what then was called "America's Weekly Magazine," *Collier's*. They bought it on sight, and ran it to good reader reception.

LESSON:

Don't overlook your everyday dealings and surroundings. It may be prosaic to you from familiarity, but may be of interest to those who don't see it.

14.8

Ben Franklin said, "A word to the wise is enough." Not every professional, outstanding in other areas, has the *natural* talents or the *acquired* skills to become a good author. For those, the accomplishment of a superior report or paper is sufficient. Only *you* can know if you have that natural *talent* or cultivated *skill* and pride to be an author. The next two chapters are available for everyone, but they are targeted at you.

> GREAT SPIRITS HAVE ALWAYS ENCOUNTERED VIOLENT
> OPPOSITION FROM MEDIOCRE MINDS.
>
> —ALBERT EINSTEIN[1]

NOTE

1. 1879–1955.

15

ARTICLES

15.0 MOTIVATION

If you want to be paid, you write reports. If you want to spread knowledge, or disseminate *your* latest findings, you write journal papers. And if you want to advance your status, you realize your professional expression through writing articles.

By now you have read, understood, and *accomplished* the material in the preceding twelve chapters. You should have *mastered* with ease, writing reports and journal papers. Your next step is to produce good *publishable* articles for magazines, newspapers, and brochures (sec. 15.6).

15.0.1

To dip into a bit of "alliteration's artful aid,"[1] your motivation for publishing will come from among these six: progress, professional pride, prevention of punishment, prestige, pleasure, and payment.

15.1 READERS

Article readership is voluntary. To any reader, his most precious possession is his *time* of life. For him to spend some of that irreplaceable commodity on an article you have written (sec. 15.5.2), you must first win a silent, static competition for his time, then you must reward him with material of *lasting value*.

A published article must sell itself in its own interest area. People who read *Mechanical Engineering* won't pick up *Seventeen*; *Field and Stream* enthusiasts won't read *Playboy*.

Most people read magazines to pass idle time in professional waiting rooms, during travel, or at ease with no planned agenda. A reader will browse the pages until a headline, a subhead statement, a pictorial graphic, or a byline name catches his attention. Once he is attracted, your lead paragraph will *hook* him. Then either he reads to the end, or gets bored and moves on. A small percentage of individual readers seek specific articles in specific magazines for specific material; even then your article is a *self-sell* item.

15.2 ARTICLE ATTRIBUTES

You start with your finished product from the first twelve chapters. Keep to your original reason for writing (secs. 1.4 and 2.1.2), and to your now-expanded target audience (secs. 1.5 and 2.1.3). Now, instead of an individual or a small number of readers, your expanded audience becomes a selected population segment that is determined by the readership demographic of your selected publication. Weave in these things:

15.2.1 ARTFUL

Avoid the mundane; make it artfully attractive.

15.2.11 APPEALING

Make your up-top items be what he wants to see and read.

15.2.12 ARRESTING

Once you've caught his eye, keep his interest by covering material he *needs*.

15.2.13 APPETIZING

Offer him something he *wants* to read.

15.2.14 ABSORBING

Keep his interest from waning by including *more* about the subject.

15.2.2 ACCEPTABLE

Unless you intend to be spectacular, don't include anything to raise his doubts or pique his ire.

15.2.21 AUTHORITATIVE

If you firmly believe it, cause him to have confidence in your judgment (sec. 15.2.8).

15.2.22 ASSURING

Make him comfortable with it.

15.2.3 ACCESSIBLE

Lay it all out clearly; don't make him ponder or dig for your meaning, he may not.

15.2.31 AVAILABLE

Include all the material you can find, and add sources where he can go to find more.

15.2.4 ARTICULATE

Never patronize or write down to an intelligent reader.

15.2.41 ABUNDANTLY CLEAR

Avoid circumlocution and ambiguity; don't leave him in doubt.

15.2.5 ACCURATE

Steer clear of nebulous statements; be sure everything you say is as it is believed to be.

15.2.6 AFFLUENT

Without padding or bloating, give an *abundance* of even peripheral information if it contributes to the presentation and does not get in the way.

15.2.7 ANECDOTAL

Small incidents or recurrences, and even humor (in good taste), to an optimum amount, can humanize your narrative without detracting from the article's thrust. It relaxes the reader.

15.2.8 AFFIRMATIVE

Put all your material into a positive light, to avoid reader doubts (sec. 15.2.22).

15.2.9 ACKNOWLEDGMENTS

Credit (even the smallest items) with correct and correctly spelled names, everyone who provided you with any fact or help. It makes them feel good, and you might want to go back there again.

15.2.10 AFTERTHOUGHTS

None. No "this just in" sort of thing after you've sent off the manuscript. Avoid what the French call *arrière-pensées*.

15.3 ACTIVATION

Articles of limited space cannot encompass all aspects of a subject. Along with an assignment, an editor will give you a deadline and a maximum word count. Be sure you include everything reasonable to cover your subject thoroughly. Don't deluge your reader's mind with material you can't explain fully in the allotted space, and don't drop in an item without telling your reader how it fits, or why you've included it.

15.3.1

An *overall* (general) article yields broad area coverage. It is a perspicacious, panoramic overview of a subject or situation. Its text is generally superficial with little depth.

15.3.2

A *specific* (detailed) article can be as long or longer than an overall article. It is focused on a small section of an overall article's coverage, but goes into considerable *depth*. Example, an *overall* article could be about North Atlantic weather, explaining all aspects, mentioning hurricanes as part of it. A *specific* article, a completely different item, would just be about hurricanes, in full depth, detailing every aspect of them. You do not need both for a single story, nor an overall article from which to spin off a specific article. Each is a freestanding entity.

15.3.3

A *person piece* is a special case of a specific article (sec. 15.3.2). It deals with an individual and his current, particular activity or surroundings.

15.3.31

An *interview*. The *subject* is the focus, including how that person fits into it.

15.3.32

A *profile*. The *person* is the focus, then how he deals with that subject.

15.4 APPROACH AND ADVANCEMENT

Stop. For the moment, please return to sections 4.2 and 4.3. Reread them carefully, and give that combined advice maximum credence. They lay out in clearly detailed order the foundation you must construct for your article. Then:

15.4.1

Set the article's scope, or pose a rhetorical question early in the lead paragraph, for example,

> *Should nanotechnology be taught in high school?*
> *Can a second Chesapeake Bay Bridge system relieve north-south traffic congestion?*

15.4.2

Completely define the focal venue and situation to put your reader into your article.

15.4.3

Create a catchy or provocative title.

> *The Seven Constructed Wonders of New York City*
> *Air Photo Interpretation Finds Oil*
> *Girder Holds Up Des Moines Bank*

15.4.4

Add a subtitle, partially to expand on the title, partially to entice.

> *New findings may lead to allergy abatement.*
> *BACnet enables greater operational control.*

15.4.5

Start with a solid lead line.

All cities of the world are alike, except Venice.

Penn's captain-elect played the entire PU-CU game on the Big Red team.

15.4.6

Salt in occasional paragraph headings to relieve voluminous continuous text.

Equipment is portable.
Radiation is contained.

15.4.7 SECURITY

Unless you have a specific mandate, use discretion. Avoid conflict with special subject concerns about property, propriety, policy, and politics. They can only lead to difficulty, present and future.

15.4.8 GRAPHICS

In an article, graphics are mainly decorative, for reader comfort. Often they illustrate a part of the text (sec. 5.3). Occasionally they catch a reader's eye. In an article I wrote about a runway expansion at New York's LaGuardia Airport, I used a page-top photo of the field that showed the subject runway from the page's upper right corner, angling down and left to the lead line, to draw the reader's eye to start reading.

15.4.81 CAPTIONS

A picture can be worth many words *only* if the caption tells your reader what the graphic is, what specifically is its focal point, and how it relates to the article.

15.4.9 ENDING

Summarize the high points of the article *briefly* in the last paragraph, but (depending again on the reason and the reader), you do not necessarily need a conclusion. Close with ease, but leave nothing dangling nor open-ended.

15.5 RESPONSIBILITIES

As an article author, you undertake five *responsibilities* when you publish a piece.

15.5.1 TO THE *SUBJECT*

Show it in its most understandable and best light unless you are writing a negative article for a specific cause, then treat it fairly.

15.5.2 TO THE *READER*

You owe him something of lasting value (sec. 15.1) in return for his spending valuable time on your written product.

15.5.3 TO THE *PROFESSION*

As a member of your chosen profession (also a peripheral member of the literary profession), you should release for publication only material that is true, accurate, precise, and positive, to bring credit to both those professions. Never take advantage of another's misfortunes.

15.5.4 TO THE *LANGUAGE*

American English is your national and your professional idiom. Adhere to its rules and irregularities. Use your language well as your professional expression vehicle. Protect it from pop-culture shortcuts.

15.5.5 TO *YOURSELF*

Take pride in *your* professional attainment and stature, and in the written effort you have produced for the world to see, then think well of *you*. To do well for all your responsibilities, use the checklist in appendix 15.5.5.

15.6 MARKETING

At the start it is better to write your article when the idea comes to you, and you know or can find all the necessary information. Later, *after* some publication experience, do the writing and the marketing simultaneously.

15.6.1 PROCEDURE

The *Writer's Digest* magazine and the Internet can give you sources of information by categories of publication titles, names of contacts, e-mail and postal addresses, and phone numbers. Make a list of possibilities in your order of choice. Then compose a query sheet format.

Address it properly to a specific *person* by name. At the top, briefly tell what your offering is about; why it would benefit that magazine's readers; how many words; what illustrations; and when it will be ready. Add the proposed title, subheading, and your lead paragraph. Burn it to CD and say you can also send by hard copy or by e-mail. Get all that onto *one* page, then on a separate sheet, add your brief professional biography. Add tear sheets of one or two of your previous publications, but no more. If you have none yet, don't mention that. We all have to start somewhere.

If the magazine has specific guidelines for submitting manuscripts, follow them.

If you want to include a complete manuscript in hard copy, remember most publications demand you enclose a self-addressed, stamped envelope if you want it back. If you are *really* confident, don't bother.

Follow your entry with a phone call to remind your contact if goes beyond ten days. This may trigger a positive response, or at least build good will.

There may be several magazines for which your effort would be suitable. If your first choice doesn't take, try the others in order, but only *one at a time*. Perish the thought that you broadside four magazines simultaneously and they all accept!

15.6.2

If you receive an acceptance, contact the sender immediately to accept an editor's suggestion on adds, deletions, or revisions; a deadline; and a *stipend*. You want to be paid for your work.

15.6.3 SOME CAUTIONS

You may have a great idea or a great article, but the timing or climate may be wrong. Two old personal examples come to mind: I was invited to visit the Soviet Institute of Construction and Architecture. During evening hours I was entertained and intrigued with the number and quality of the concerts in Gorky Park. It made a great article, but my editor and friend at a national magazine turned it down. He told me to save my time trying elsewhere because "no magazine today will ever take a piece that shows Russians as human beings."

From a press trip to Israel, I reviewed a historic meeting presentation about a joint study among civil engineers from Israel, Jordan, and Syria. Among them cooperatively they devised a plan to use the Jordan and Yarmuk Rivers in a project to provide adequate water supply and drainage control for the entire region. It never got off the ground! Although the engineers were amiable, the politicos were not. No commercial magazine would touch that political hot coal, but the *Cornell Alumni News* did, to good reader comment.

All that adds to the actual article anecdotes in section 14.4.

IF YOU CAN FIND A WAY TO WRITE IT,
YOU CAN FIND A WAY TO SELL IT.

NOTE

1. "Who often, but without success, has prayed for apt alliteration's artful aid." Charles Churchill (1743–64), *The Prophecy of Famine* (London: G. Kearsly, 1763), 5.

16

BOOKS

16.0

"Of all needs a book has, the chief need is that it be readable," wrote Anthony Trollope, in his autobiography.[1] To that, Henry James added, "be interesting"[2] (app. 16.0).

The way to write a good book is to *read* good books; many of them, over a long period. But don't read them as most people do, just by absorbing or noting the content material. Pay close attention to the author's style, the way he lays out his ideas, the way he describes his entities, the logic he uses to prove his points, and the manner in which he reaches you. (app 16.0.1)

Not every writer is able to author a book. First you must have something of *lasting value* to say, something useful, or something to improve the quality of life.

16.1

Professional expression is not confecting novels. Yes, it can be colorful (sec. 5.3), but it must be true (and proved), and be sincere. You should have *experienced* the subject *yourself* to be able to impart that confidence to your reader. The old cliché goes, "You write best what you know best."

16.2

If you have a new idea, or something in addition to or instead of something extant, talk it over with a person you respect and *trust*, to gain encouragement. Then think it through, on paper. If you decide it is *indeed* worthy of your effort (it isn't easy), then you might want to approach a *reputable* publisher. Most good publishing houses know the trade, know their profession, and know the market. If you are turned down in your first attempt, try another, and another. Often, for whatever reason, a publisher will nix your offer by writing, "It is not right for us at this time." That is not an outright rejection, and at some reasonable future date you might want to try them again. The climate may have changed *then*, and your idea perhaps will be just what they seek.

16.3

Don't be discouraged if your early attempts don't catch fire; if your offering is solid, it will. You might be looking in the wrong place. Some writers work on an idea for years (sec. 16.7) until the right editors and the right time coincide—propinquity.

Do not abandon your quest without a test run, as some suitable adjustment may make it work. Yet there is always this realistic possibility: you might be chasing a rainbow, and your project may not ever take off. You'll never know unless you try; but *do* know when to stop whipping a dead horse.

CAUTION:

Eschew publishing on your own via a "vanity press"; that can become a highly expensive proposition. The time, energy, and money you must invest, never can be requited. That route lacks the solid organization you need to promote, store, distribute universally, and manage the business required for your book to reach the audience you hope to serve. Established publishers do know something you don't.

16.4 GROUND ZERO

Start with a valid premise as your cynosure. To sketch, draft, and write your text, follow the suggestions, directions, and advice you worked through in the fifteen preceding chapters here. You need no further spelling out about writing. Thus the emphasis of this chapter is on *marketing* your project.

Toward that end I offer you the experience I've gleaned from successfully having written two complete books, compiled and annotated three anthologies, and edited ninety-three volumes in print.

Reliable publishers constantly seek something innovative, something better than the extant. They are always willing to *have a look* at the next suggestion.

All publishers have their own proposal formats. They may differ in details (where the devil hides), or in the order of answer presentation, but ostensibly they seek from you enough information to enable their editors to make an *informed* evaluation of your offering.

16.5

Proposal formats are proprietary for each house, but since they all pose much of the same inquiry, I *can* offer you the form for this publisher, Momentum Press, as a prototype for your proposal preparation:

PROPOSAL GUIDELINE

We are always eager to receive and evaluate proposals for new books and other types of professional publications, whether print or electronic. These guidelines have been prepared to help you write a proposal that presents accurately your ideas for our consideration. They should encourage you to think through your project carefully and determine an effective framework for its completion. Thought and care at this early stage will be rewarded amply as your book develops. If your proposal follows our suggestions, the evaluation process will be efficient and accurate.

TEXT AND CONTENT

Briefly describe your project, its rationale, and its approach. Include a working title that clearly defines the subject area. Indicate how your project relates to the field and what essential features it will contain. Please also indicate your qualifications for writing or otherwise managing and editing this book or project.

We need to know as much as possible about your topic coverage and organization; a detailed table of contents, series outline, or component outline (for electronic products) should be prepared carefully. Please list individual chapters, or monographic titles (if a series), or component names (for online content). If possible, provide a paragraph or two about the content of major chapters, sections, or components.

Identify the current leading books or similar informational products in this field and briefly discuss their strengths and weaknesses. Does your project compete with or complement those other publications? If possible, please provide complete publication information for each competing/related book, series, or informational product, including author (or editor or company name), title, page count (if applicable), and if available, the ISBN number or other identifier by which this publication may be found (e.g., URL address).

MARKET AND READERSHIP

- For whom is the book or product intended?

    Professionals in the field?
    Researchers and scholars?
    Advanced undergraduate and/or graduate students?

How will readers make use of the book or informational product? What tasks will it help them accomplish? What societies would the professional or students typically belong to? What magazines or journals would those professionals or students read? Please give careful consideration to who will need the book or otherwise make use of this informational product and how your customers will learn of its existence. Do you have any idea about the actual size of the market (e.g., any statistics you can provide or demographic information)?

- What is the level of presentation of the subject matter (e.g., what are the necessary educational or professional prerequisites for readers or users of your book or informational product)?

- If this book or informational product is going to be used primarily as a textbook, in what courses would it be used? Can you identify schools where such courses are taught? Also, do you intend to provide any ancillary documents or tools to use with your textbook, such as a Solutions Manual, Study Guide, Workbook, or accompanying software of any kind?

STATUS OF THE PROJECT

- What is your schedule for completing the book or project? What portion is already finished? If possible, please send a sample chapter, section, or operational component (e.g., for electronic products).

- If this project is to be a print book or other print publication, please estimate the length of your manuscript, either in pages (double spaced on 8½ × 11 white paper or the equivalent standard size paper if outside the United States), or in number of words. (In general, a page of all text writing, on a standard 8½ × 11 sheet, created in 11 or 12 point type contains about 250 to 300 words in a double-space or space-and-a-half format—illustrations, tables and so on will reduce that number of words, accordingly. To estimate the number of words in a published book, a page of text in a standard 7 inch by 10 inch book contains about 470 words.) Also, please estimate the number of illustrations (line drawings and photographs). To help you make this estimate quickly, try comparing your proposed book to other similar books with which you are familiar. You can also estimate the length based on the number of proposed chapters and the likely average length of each chapter. For estimating illustrations, you can make a rough determination by assuming an average number of illustrations per page, and then multiplying by the estimated page count.

- If this project is to be an electronic publication, please try to estimate the total file size in megabytes (MB) or gigabytes (GB) needed to contain the entire product. Also, if possible, please provide the file type and operating platform which will be used to create this product and what will be needed to host it online (if applicable).

- For both print and electronic publications, please indicate the software/program that you will be using to create the text portion, and the file format in which you will store your manuscript files. Likewise, please provide the same information for the illustration and photographic files, as well. Also, if you are planning to submit your print publication as "camera-ready" (i.e., with the pages laid out exactly as they would appear in the final, published book), please let us know. In the event that you are planning to prepare your book as a camera-ready publication, we will need to see sample pages at the earliest possible point, preferably in printer-ready, PDF file format.

ABOUT THE AUTHOR

- We would appreciate a current curriculum vitae that outlines your education, professional memberships, current position, and previous publications. In addition, if you would care to do so, you may submit copies of recent articles or journal papers you have published, or online projects that you have created or helped to develop.

REVIEW PROCESS

- Your proposal and any supporting material will be read by me, and discussed with colleagues and advisory editors, who observe strict confidentiality. Their comments and suggestions will be communicated to you.

- We would also like to have the proposal reviewed by reviewers you suggest. Please suggest two or three reviewers who you feel are qualified to review your proposal.

We appreciate your attention to preparation of this preliminary information.

If you have any concerns or questions, please contact us directly. When complete, please forward your proposal to:

Specific name
Proper publishing house
Address

16.6 TEN LESSONS FROM EXPERIENCE

For the most part, people who write books have assertive personalities. Otherwise they would not make the assuring positive statements required in technical texts. There is a distinct difference between *assertive* and *aggressive*. Publishers tend to avoid aggressive writers for obvious reasons. Writers lacking the confidence concomitant with assertiveness may never see their work in print.

Fortunate are those who have put together from experience the makings of a valid book that may long lie fallow among their files until some person or occasion brings it out. Eight more make my point, but two cases are worth noting about stored embryonic good books:

16.6.1 READY RESOURCES

Harvey V. Debo's valedictory as a senior project manager for Turner Construction Co. was the complete rebuilding of New York's Penn Station without interrupting the voluminous daily railroad traffic. Often visiting that job, we rejuvenated our friendship that began when I worked for him on the construction of the flight test hangars at the Rome, NY, Army Air Base.

During our later brief lunches at Penn Station, he would recount his vast and varied construction experiences over this past half century. Realizing immediately that there was a wealth of techniques and innovative solutions for unexpected field obstacles to fill a practical construction guide, I suggested that to him. He responded readily and positively, saying, toward that end, he had kept scores of notes, waiting for such a chance.

I proceeded to conjure a proposal and secured a contract for him. As I informed him, he handed me a *finished* draft manuscript, all handwritten on legal-size, lined yellow paper. It needed considerable editing, but I would not alter his plain, experienced, professional voice. Unfortunately, he died too soon.

To complete a producible script, I persuaded Cornell classmate Leo Diamant, PE (who had just completed project managing the concrete base for NASA's Cape Kennedy launch pad), to wrap the Debo script while preserving the field flavor. Later Leo told me he had always wanted to do just such a book, but never felt the needed spark. Harvey's guide enjoyed a successful life well into its second edition.

16.6.2 READY WHEN YOU ARE

H. Michael Newman, the man responsible for Cornell University's electronic controls, and author of *Direct Digital Control of Building Systems*,[3] produced a valid volume about *BACnet* (a data communication protocol for building automation and control networks). He was able to do that just by copyediting his daily job log.

16.6.3 MATURATION

The content of this very book, *Professional Expression*, was conceived, outlined, and roughed out during my first teaching semester at New York's Cooper Union, fall 1965. As time has trudged on since, the process has been enlarged, diminished, and modified to keep the lectures current, but the basic premises have remained solidly unchanged. Only superficial details and examples have been refined to the point where I felt in 2008 that the material and I both were ready to publish; and *this* is it.

16.6.4 NEW

Dr. Uzi Mann, professor of chemical engineering at Texas Tech University, developed a new methodology to design and analyze chemical reactors. Recognizing that this methodology provided design capabilities that are *not* taught in the undergraduate curriculum, he developed a new course during his sabbatical at Cornell University in 1993. Wishing to foster the first working text on the methodology from the source, I asked him to write it. He welcomed the idea, but would submit the manuscript to the publisher *only* when it was free of flaws and solidly ready. A decade and a half later, in 2008, the methodology became a published book. "Patience and time."[4]

16.6.5 SERVING ALL LEVELS

Dr. Fu Hua Chen was the acknowledged world's authority on expansive soils. He also had been the chief engineer on the Burma Road, the Tibet Highway, and the Ho Chi Minh Trail. A practitioner and a scholar in both the United States and China, he wrote extensively on his expertise and achievements, but all were expressed on the highly technical and intellectual levels of his profession. Under it all he always wanted to author a simple (but not simplistic) guide to *demystify* the elite province of soils and foundation engineering. Publishers and peers offered him no encouragement.

Determined to write that after his higher level publications, he enlisted the help of two top practicing professors to produce the volume he desired before "the rice is cooked" (as he put it from his wealth of Chinese proverbs). As his editor, I was later able to get him an advance copy of his book-in-print before he died peacefully at age eighty-seven.

16.6.6 ADVANCEMENT

B. Austin Barry, FSC, PE, was a Christian Brother and a professor at Manhattan College. He wrote a serviceable surveying guide that went into a second edition. Motivated by the success of that work, and armed with both the mental agility and an easy turn of phrase, he made a huge leap forward by authoring *Errors in Practical Measurement in Science, Engineering, and Technology*, a universal book about statistical measurement and error theory. He unwrapped reliability of measurements, and of repeating them, propagation of errors in computing, and he related *accuracy* to *precision*. None of that material had ever been offered before to readers in such comprehensive terms. In a huge leap of good judgment, John Wiley & Sons published it successfully.

16.6.7 PROTOCOL JUDGMENT

John J. P. Krol, PE, Esq. wrote an informative volume, *Construction Contract Law;* as an editor, I asked it be produced in gold-lettered red cloth hard binding to appear as the law book indeed it is. It does not deal with defining the law, in *legalese*; rather it tells a contractor what, within the law, he *can do for himself*, and what he *cannot*. It tells when he *needs* an attorney, and what he can expect from one. Since any novice seeking legal advice or assistance would have to pay about $150 for a first hour's visit, it would seem reasonable for this reusable preemptive tool to be priced at $95, pricey in 1993, but less than visit #1.

With all that going for it, over its first *fifteen* years on the market it has sold fewer than 1,000 copies, but remains in the catalogue. Why?

Because the publisher is major, thus this unique product became merely one of the many thousands of titles it publishes in any year. The item is entered into its current catalog where anyone can find it and buy it. But without advertising how will a potential buyer know it exists, or what its benefits are?

A smaller house would make broadside mailings to potential readers in various interest areas, and actively promote organization reviews.

It is different if you have a *hot subject*, or a *known name*. Either of those would work well with a major publisher for the benefit of the house, the author, and the interested public. For the average, or first-time author, you might find more attention at a smaller publisher. Choose carefully.

16.6.8 IMPROVE THE PRESENTATION ARRANGEMENT

Section 14.6 emphasized the importance of choosing carefully your material arrangement layout. The order in which you present your text shapes the reader's mental picture of your ideas, thus determining his view of your book's strength.

Two bright young southern university professors generated the good notion of using the literature to tie closer the engineers who create the world around us, and the humans who enjoy the benefits of their work. Their first submission of that sterling idea was a random collection of good, meaningful fragments culled from the literary larder. That grab bag lot would not attract an audience, much less retain one. The first revision listed the entries in alphabetical order of the author's names: no visible subject relationships. The second revision put the selections into chronological order: no significant interrelationships. Finally, steered to it, they set the only logical arrangement to prove their point: a set of category chapters, into which all selections could be sorted (engineering and literature, art, philosophy, and history; then closing with an educated guess look into developing nations and into the future). Each chapter then began with a pile-cap commentary, tying the separate entries to each other. That was the format in which the volume went to press.

16.6.9 RESOURCEFULNESS

A *dictionary* is a broad area coverage, alphabetical compendium of words. That generally includes pronunciations, definitions, uses, and sometimes origins. A *glossary* is much the same, *except* it is confined to words of a specific interest area. Both are *lexicons*.

In the late 1970s, J. Stewart Stein, an architect, and Fellow of the Construction Specifications Institute (CSI), offered my then publisher a comprehensive glossary of construction and related words. Tactfully I demurred because there currently were several volumes of that genre. He had a worthy project but needed a new approach.

Lexicons generally deal with terms by listing all possible definition variations (1, 2, and up) in their separate contexts, sometimes in the order of their most frequent use. You have to plow through the full listing to find the specific one you seek.

Resourcefully, Stein employed the CSI-delineated sixteen separate areas of construction as chapter headings. Within each, he listed and defined the words alphabetically, once, as they applied to that dedicated division only. The same individual word might appear in several division chapters where they would fit, and not in categories where they did not. Ergo, if you sought a word in one connection, it would be defined in that context only. Since his 1980 better mousetrap, Stein's segmented glossary has been successful and useful.

16.6.10 FORMAT FOLLOWING FUNCTION

Karl F. Schmid, PE, has been Assistant Commissioner of Buildings for the City of New York, and earlier, the owners' representative in the construction of buildings and facilities for Cornell Weill Medical College, then Baruch College. He was also a paratroop engineer officer in Vietnam. During the years he had been alpha dog in those operations, he managed them from the ground, among the force. He realized whether it be a factory procedure, a ship, a shipyard, or a field force, it was the foremen, the sergeants, the "straw bosses," on the line or in the pits, who *got the job done*. Regardless of the project, the spot leadership was essentially the same everywhere.

Schmid organized and wrote a short, potent field leadership manual for foremen and sergeants, in their own idiom and voice. The text was displayed with some graphics to be an inexpensive pocket-sized paperback, much as the *Strunk and White* writer's style guide. Several publishers turned it down.

The problem was not in the book's style nor format, it was in publisher inflexibility. Too many established houses become myopic toward the potential market, demurring because a different kind of offering does not fit their procedural protocol patterns. Recently this publisher took a reasonable gamble. Schmid's manual, usefully, will be in the field in 2009.

16.7 A WORD OF CAUTION

You have finished your last page proof review and released it all to production. After that, there is little you can do for or about your book-to-be. The paramount consideration for your cautious concern now is the dust jacket, or the cover graphic. That is the first and only face your potential buying public will see: it *has* to be good. Meet with your editor *before* he signs off on that key up-front item to be sure the artwork truly projects the most accurate image of *your* written product.

My outstanding prime instance of that situation occurred in 1976, when Dr. John B. Scalzi, with Walter Podolny, cowrote the first (and landmark) technical treatise on *Cable Stayed Bridges*. With great artistic taste, but no awareness, someone in authority selected a dramatic photo of the Golden Gate Bridge at sunset for the dust jacket. Fortunately I caught and stopped the approved copy just as it was headed for the press. The Golden Gate is a cable *suspension* bridge, as different from a cable *stayed* bridge as a Great Dane is a different dog from a poodle. Had that photo cover gone ahead, it would have been a prestigious embarrassment for Wiley, and for the authors. "After they lock the barn door, it's too late to steal the horse."[5]

> A GOOD BOOK IS THE BEST OF FRIENDS,
> THE SAME TODAY AND FOREVER.
>
> —Martin Farquar Tupper[6]

NOTES

1. Anthony Trollope (1815–82), *An Autobiography* (London: Trollope Society, 1999; originally published 1883), chap. 19.
2. Henry James (1843–1916), "The Awkward Age," in *The Art of Criticism*, ed. William Veeder and Susan M. Griffin (Chicago: University of Chicago Press, 1986; originally published 1908), 310.
3. H. Michael Newman, *Direct Digital Control of Building Systems* (New York: John Wiley & Sons, 1994).
4. Russian General M. I. Kutusov, in *War and Peace*, by Leo Tolstoy (Moscow, Russkii Vestnik, 1869).
5. Author unknown.
6. Martin Farquar Tupper (1810–79), *Proverbial Philosophy of Truth in Things False* (of Reading) (Auburn, NY: Alden & Markham, 1848).

17
CRITIQUE, DISCUSSION, REBUTTAL

17.0 RESPONSES OR REACTIONS

Responses or reactions to communications received direct or via publications indicate your own differences with the actual or opinion material on the record; they are *not* to impugn any person.

A captious PhD professor I knew, assiduously attended meeting lectures and read copious technical journals, but never was able to get himself together enough to write a paper or an article of his own. Yet he backed into print relentlessly, by writing long and detailed discussions dissecting other's efforts.

Once I gave a well-deserved "A" to a capable student. She later applied *unsuccessfully* for a staff spot at a good magazine. Subsequently, on her own, she wrote a monthly letter-to-the-editor critiquing (up or down) its current cover story. After four months of her letters appearing in consecutive issues, the editor called her in, took her to lunch, and offered her a post. The editor was somewhat embarrassed when the former student told the editor she had turned her down just five months earlier.

17.1 ELIGIBILITY

To write a critique, discussion, or rebuttal (CDR), you have to meet *all* three of these (3 for 3) prerequisites, then have a clean (0 for 6) sweep of section 17.3.

17.1.1

You must separate opinion from fact. Because you believe something is so, that does not necessarily make it so; be careful.

17.1.2

Do not criticize unless you have something better to offer (app. 17.1.2). Just to say something is bad is not enough; propose a superior suggestion, and document it.

17.1.3

Apply all the rules for minimum standard of reader acceptability (sec. 1.3); especially make your communication commensurate in size with the *value* of the message it carries (sec. 1.3.3 and chap. 11).

17.2 DO WRITE

These are easy, and far more positive. Write a CDR for any *one* of these *eight* reasons (1 of 8). You may find more. Your chances of having all eight are 1:40,320. Write if you see:

17.2.1 SOMETHING WRONG

It can either be obvious or obscure; but not minuscule, that burns valuable time. Here is an excellent example:

> To The Editor:
>
> The authors of "SEMICONDUCTOR MATERIALS," (RADIO ELECTRONICS JOURNAL for April 19th, pg. 67) made an error in explaining the key conduction mechanism of a P-N junction. They say:
>
> "The width of the junction decreases until it approaches the potential of the battery. The diode begins to conduct from the P side to the N. The electrons flow from negative to positive."

The last statement is incorrect. The P side produces carrier electrons that make that section positive. Conversely, the N side accepts electrons that make it negative. Electrons flow from positive to negative and carriers from negative to positive.

Reference: TRANSISTOR CIRCUITS by R. Turner (Prentice Hall, NJ).

Michael Zarro, ET, FDU

17.2.2 AN ERROR OF OMISSION

Missing material can lead a reader away from your desired conclusion, or cause him to go back to you with questions.

(On a symposium in American Society for Testing and Materials [ASTM], "Materials Research Standards.")

M. D. Morris—Although his paper is a short synopsis of the history, development, and state of the art, in its efforts at brevity, it has omitted mention of many salient points important to the practice of obtaining good, useable aerial photographs to meet today's engineering needs. Those items are detailed (up to date of publication) in my article "Aerial Photography," in the CONSULTING ENGINEER, April 26th (also not mentioned among Mr. Parker's 45 annotated references).

Omitted was mention of deterrents as: fog, clouds, atmospheric heat waves, and "blurbling"; air motions (forward speed, drift, roll, pitch, yaw, and vehicle vibrations); and the electromechanical devices to overcome them all. Also, radiation effects on film, high resolution film, and high acuity lenses are not covered.

Mr. Parker's discussion of radar omits mention that resultant photographs are pictures of the radar scope, not photographs of the ground. His opening premise that "Aerial reconnaissance some day will supplant soil augers, rock drills . . . " etc., can only be wishful thinking. Surely AR will augment or preempt, but never replace in-situ sample qualitative analysis, because only by this real, local method, can we confirm beyond doubt what we have learned about large areas from aerial reconnaissance.

17.2.3 SUPERFLUOUS MATERIAL

Marginally related material or anything not pertinent to reaching the intended objective economically only pads the word count and distracts from reaching the desired conclusion.

17.2.4 BLANKET COVERAGE

The author provides broad coverage, but does not include exceptions.

17.2.5 APPLICATIONS TO NEW IDEAS IN THE SUBJECT AREA

If you see possibilities of additional uses for the subject to enhance the working scope.

17.2.6 SPRINGBOARDS TO IDEAS IN OTHER FIELDS

What works in your interest area might well (or, with modifications) be the solution to a problem in another, unrelated field. Sir Henry Bessemer (1813–91) looked at a working Russian tea samovar and extrapolated the idea into the Bessemer steel furnace—perspicacity!

17.2.7 A WEAK CASE

If you like someone's idea and have considerably more data to substantiate it, that makes it stronger, and it makes the author feel better, and you look better.

17.2.8 A SUPERIOR EFFORT DESERVES PRAISE

People are fast to criticize, but seldom compliment something good. A pat on the back improves working conditions, and work.

Any *one* of those eight is sufficient to write a CDR, but you may use as many as you need (sec 17.2).

17.3 DO NOT WRITE

Do not write a CDR if you see any *one* of these six situations. You must have a clean sweep (0 for 6). Be gracious: even if you have all eight in section 17.2, the don'ts of 17.3 *govern*. Don't write a CDR just to:

17.3.1 CITE A PERSONAL EXPERIENCE

Do not cite one unless it truly contributes to the profession, or conflicts with "conventional wisdom."

17.3.2 ENGAGE IN PERSONALITIES

Stay with the subject. The author may have the world's worst disposition, but if his facts and figures are correct, and his premise is reasonable, leave him alone.

17.3.3 MAKE JESTS

The author's paper may contain whatever humor the subject or occasion permits, but you have no license for levity with another's work.

17.3.4 COMPOUND AN ERROR

Be absolutely certain of your facts. If the author goes off the edge, don't let yourself go with him.

17.3.5 CORRECT TYPOS OR SOLECISMS

That is the copyeditor's job.

17.3.6 FAULT THE AUTHOR'S STYLE

Don't criticize unless the entire original is impossible to understand. Style is personal to each individual.

If you have passed by all six (0 for 6) *don'ts*, and have at least one of the eight *dos*, you are good to go.

17.4 GROUND RULES FOR WRITING A CDR

17.4.1

Cite publication or report title, volume, issue number, and date, and the author's name.

17.4.2

Cite verbatim, in quotes, the section or statement against which you direct your CDR.

17.4.3

Cite the specific reason (sec. 17.2) for *your* writing.

17.4.4

State what you believe the correct statement should be and cite valid, published references to substantiate your view.

17.4.5

Remember the author has his closure; be certain *you* are correct.

17.5 LETTERS TO THE EDITOR

Letters to the editor are an informal, special case for CDR, requiring you to comply with sections 17.1, 17.2, 17.3, but not 17.4. Apply this modification of section 17.4:

17.5.1

Cite issue date, subject, title, and author.

17.5.2

Cite verbatim the questionable idea or specific quote.

17.5.3

State what you believe to be correct, then substantiate it with valid references.

17.5.4

Make it brief, make it light, be witty if it's good wit, and know when to put your period.

> CRITICISM COMES EASIER THAN CRAFTSMANSHIP
>
> —Pliny the Elder[1]

NOTE

1. *Natural History*, 71 AD, (from Zeuxis c. 400 BC.)

18

LETTERS

18.0 LETTER WRITING

Letter writing is an endangered art. People tend not to write real letters anymore. Thus, unfortunately, the common skill of writing a complete letter, professional or personal, is dying out: soon to be extinct, along with the mimeograph and the 78 rpm record, to be gone, except from those among us who care about interpersonal, human communication.

The contemporary jargon of e-mail, text messaging, or worse, twitter, so taken for granted by today's teenagers as normality, also has pervaded their daily domestic discourse. Some examples of our future generation's conversation: TMI, *too much information*; totes, *totally*; you rock, *you are great*; google, *to search for information on the Internet*; LOL, *laughing out loud*. All that ingrained *before* they get to college will have diminished their adult communication ability, and consequently their professional expression.

Combating that degeneration certainly is not as dramatic as saving the environment, or salvaging the mother tongue, but it has become a definite concern. This enlists every currently working professional, paraprofessional, and technician to take the effort to write better letters.

18.1

Years ago I gave a letter writing lecture at an American Society of Civil Engineers (ASCE) meeting in the form of a model letter: it was later included in its newsletter. I reuse it here now because nothing has changed; I've already done the drill; and it all remains what I advocate.

YOUR LETTERHEAD

Today's Date

Greetings Letter Writer:

"At hand is your good missive of the 7th instant, and the undersigned hastens to respond to the technological points of inquiry its writer had included therein."

No professional in his right mind would walk into a person's office and address him in such a mode of mid-Victorian verbiage. Then why write it? Even today, too many people feel compelled to impose such drivel into their commercial correspondence.

Actually on arriving for a personal visit, you might instead say: *"Your December 7 letter raised some technical questions that I'd like to answer now."* Such a straight script is also writable. The difference is dramatic!

Since you cannot be everywhere and still be in the office getting your work done, you must write letters. They then become your personal representatives; thus you cannot afford to let even one go out flawed, sloppy, or incomplete. Each is you, as if talking face-to-face to your receiver. Ergo, you should write a statement no differently from the way you would speak it normally to that person.

Forget the archaic idea that all technical prose must be in the passive voice, in the third person impersonal. That notion went out with spats and celluloid collars. Mark the reader, your personal contact, as "you," and the writer as "I" or "We," depending upon your posture or position in the subject at hand. And you may not switch them in midnote: None of "*I* think *you* should know how *we* feel about this." The lone exception where I and we *can* be used in the same letter is when you write about a collaboration between you and your receiver, for example, "With you as administrator and I as engineer, we shall be able to . . . ."

Contrary to conventional practice, I recommend you do not begin your letters with the prosaic "Dear . . . ." That worn-thin greeting is stale. Do replace that, now, with a livelier greeting. How about "greetings" itself? Start with "Greetings Mrs. Ozman," eschewing the dreary "Dear Mrs. Ozman." The change will be refreshing to your recipient, who is sure to read on with interest. In 1940, using "Greetings," President Franklin Delano Roosevelt made all draft conscription letters a bit easier to swallow.

There are only two kinds of letters: an initial (*unsolicited*) letter (the more difficult), where *you* start the correspondence, and a *responsive* letter, where, obviously, you are answering someone else's opener.

Responsive letters are easier because your correspondent, by writing first, has made himself a captive audience for you. You only need begin with some reference to his initial request, then proceed to answer as many (or all if you can, or will) questions in the original, to whatever extent you believe efficient. Avoid deadhead (redundant) wording, quote your information sources, and close as warmly as appropriate.

An initial letter is harder, because your firm (or you) might be unknown to your receiver. Thus, if you get your entire message onto one page, you stand a far better chance of being read. A busy executive generally will not read a multipage missive from an unfamiliar source. He might recognize your signature if he could see it, but he may never turn past the first page.

Live lead lines are your vanguard. To get past all barriers of receiver resistance, your opener must (1) get your reader's attention quickly; (2) tell what's in it for the reader if he reads it all; and (3) state what you want, up front. If you can put those three points into a single sentence like:

Your need for a flame photometer interests us.

That one-liner by itself constitutes a whole opening paragraph. It also eliminates the need for the always awkward subject line, while it draws the reader right along into the message.

Regardless of the point, instead of beginning a letter with "I," start with "You." It hits the receiver's ego target immediately, without distractions. A stilted start like *The purpose of this letter is to . . .* often turns your reader off at just that point, which the *New Yorker* magazine refers to as the "Letters We Never Finish Reading Dept."

Every letter body should display highlights and principal matters only, to substantiate the opening premise. Relegate all details to enclosures (brochures, specification sheets, instructions, data sheets, etc.). This leaves the text trimmed lean to transmit an unencumbered major idea. If he likes that, he'll read the enclosures.

Check the final copy for omissions, appearance, spelling, and other solecisms before you sign it. Release it only when you are sure it will be *you* to the receiver's eyes. And *never* write anything in the heat of anger. Cool down first.

Close your letter quickly and quietly, without cymbals and drums, but with some small humanizing note. Write nothing like "*Trusting that all the foregoing meets with your very kind approval, we beg to remain, yours faithfully . . . .*" Instead, just make your point, and put your period.

Regards.

Sincerely,
s / You the writer
Title

18.1.1 GENERAL RULES

Writing for yourself, use "I." Writing for the firm, use "we." Do not use both in the same letter. That is inconsistent and confuses the recipient. Never refer to yourself as "the writer," nor "your correspondent," nor any variations. Always refer to the recipient as "you," never as "the reader."

Even when writing from an angry position, always write in a reader-friendly tone. That can defuse what could become a hostile situation. Never let pique get the better of your better judgment.

Address your letter with the receiver's full name and professional or academic suffix initials, proper title, organization name, correct street or PO box number, and zip code. Most important, spell his name correctly. Once, my higher bid was awarded a lucrative engagement from a purchasing agent because I phoned earlier to ascertain the proper spelling of his Slavic name; all lower bids had it wrong.

Using a separate line, apart from the address block, to direct "Attention, Charles Dawes, Chief Engineer" is outmoded. Put his name above the company name in the address block.

18.2 TYPES

Not all letters are alike; each has its individual reason for being. What you write depends upon your reason for writing (sec. 1.4.1) and who the receiver is.

18.2.1 INITIAL (UNSOLICITED) LETTERS

This is where *you* begin the exchange. This point is delineated in detail in the lecture letter (sec. 18.1); emphatically all on one page, with details in enclosures. (sec. 1.5)

18.2.2 RESPONSIVE (SOLICITED) LETTERS

This is where you are the recipient of another's initial letter. You may respond as you desire. By writing first, he has made himself a captive audience because he seeks something from you. Thus you are not limited to an "all on one page" reply. But this does not mean open season on wordiness. Be as brief as you can while still being responsive to his request. Draft your missive as a letter report (sec. 9.4). After your informational material is complete, end the letter with a cordial closing (sec. 18.5).

18.2.3 PROPOSALS

A response to a request for proposal (RFP) is an enlarged special case of a solicited letter. Unless the requester provides or demands a special format, treat it as suggested in section 18.2.2. Do not detail specifics in your text. Mention them and refer to printed enclosures (specification sheets, instruction sheets, advertising flyers) or other printed matter that provides the information already published; you need not rediscover nor rewrite. Include enough material (secs. 1.5.4 and 2.1.25) to preclude his coming back to you for more information; he might not. *You* be the bidder who responds completely to his RFP.

18.2.4 UNSOLICITED PROPOSALS

This is a case of combining appropriate parts of both an initial, unsolicited letter (sec. 18.2.1) with a responsive proposal (sec. 18.2.3). You must be cautious and tactful about this because you might be privy to some information the prospect may not want to be known. Yet apparently he needs something you can provide.

Regard the sanitized, single-page example letter (fig. 18.2.4) from "Ergo It." Successfully, it led to an order for five units. The opening line is the whole first paragraph. Outstanding, it hits the target in the most concise way, by telling the prospect what you can do for him now, and it does not disclose the source of your information. Finally, it offers a personal visit, but leaves the date and time to the prospect's choosing, and it closes properly (sec. 18.5).

18.3 OPENINGS

Among other essentials, earlier I dealt with addressing your receiver, and the sample letter tells how to get the reader's attention quickly. Very early on you must tell him what he gains from reading your full message, and state what you want. Don't leave that *want* to the end; it frustrates the reader, and it may get lost in the words.

A young man once wrote me a letter telling me who he was, what he had accomplished, and what his standing was in the community. Toward the middle of the missive I thought well of him but wondered *why* he was telling me all that. Pausing, I left the rest of his letter to read "later." Not for three weeks did I have the time to read to the end, that he was applying for a job. *That* had been filled two weeks earlier.

FIGURE 18.2.4. SAMPLE UNSOLICITED PROPOSAL LETTER

ERGO IT CORPORATION
1716 CHAMPAGNE ALLEY
SAXON PARK, NY 10999

22 October 2010

Dr. Lee High, President
Valley Engineering Associates
1066 Normandy Place
Hastings-on-Hudson, NY 10706

Greetings Dr. High,

Your need for a soil consolidation test apparatus interests us.

Our company has developed and markets a completely self-contained bench model, capable of sustaining the high, consistent pressures your work would require. Our ERGO unit is the result of seven years of research and production effort. Depending upon accessories, it costs about $5,000 f.o.b. plant, and can be shipped within ten days of receipt of firm order. Weighing 150 pounds, our unit is 12" x 16" at base, 20" high, and requires no external utility connections. It also goes packed in a finished hardwood case for lab mounting or field portability.

In addition to the advantages already mentioned, we believe our unit superior to competitive compressed air, electric, or dead-load models. There is no possibility of losing the load by losing the air pressure; no failure from blackout; no shock on loading weight increments; no heavy weights to move. Uniquely, ERGO will give you an initial "no load" reading.

We have manufactured these units for the last four of our thirty business years. Many colleges, government laboratories, and consulting engineers are pleased with their ERGO consolidation units. We believe you will be too. May we visit to discuss it further? Our representative will call early next week to arrange a time convenient to you. Regards.

Yours truly,

s/ A. Thom McEnergy, P.E.
Vice President
ERGO IT CORPORATION

ATM/emg
Enclosures: 3
    Brochure
    Specification Sheet
    Instructions

18.4 THE MESSAGE

The message should contain an expansion of your opening attention getter. Then write what you can do for him, something about yourself or your firm. Remember, first find out about your receiver, then write your letter at his knowledge datum. List your ideas in logical order, and close it cordially.

18.5 CLOSING

Your last paragraph should include a brief reiteration of the most important key thought of the message. Add an appropriate short humanizing note like "Good you've had some needed rain this late in the summer." "Hope your son enjoys his first year of college." "Hope you had a good trip to Tasmania." Unless your relationship is close, avoid things like "Trusting Vivian is feeling better these days."

President Franklin Delano Roosevelt, an erudite writer, generally would include one "new" word in each of his letters, to cause his reader to use the dictionary, and expand his horizon (app. 18.5.1 and 18.5.2). Only one though, so not to engender impatience. Also, allegedly he would include one deliberate misspelling that he would then correct by hand, to let the reader know he had read it all before signing.

Finally, always make the last sentence some form of "regards." Precede it with "best," or "kindest," or "warm," according to the degree of your acquaintance with the receiver. If it is normal, or cordial, that may make it so; if it is not, "regards" is defusing.

Before you sign it personally, check carefully to be sure that it is understandable, convincing, brief, complete, sincere, and all enclosures are in. Avoid a wordy or dramatic Tchaikovsky finale.

CAUTION:

There are some who deliberately omit one of their referenced enclosures. A week or so later they send it on with a short transmittal like, "My assistant inadvertently omitted this enclosure from my July 29 letter." That is a poor move. Instead of being the intended "nudge" or reminder device, such a transparent note only steers the reader to think, if your firm can let a small thing like this slip by, you might well omit a gear from the shipped product.

18.6 JOB APPLICATION LETTERS

Such letters could be regarded as a form of proposal letter: (1) *solicited*, if you are answering an advertisement, or (2) *unsolicited*, if you are seeking.

Find the correct name and title of the proper addressee. He could be the chief of your desired section, the personnel director, or the head of Human Resources. Then target your letter to *him*.

Starting with "Greetings": (1) "Your advertisement in (publication and date) invites this reply, etc." or (2) "You might be interested in this job application for, etc." Then follow with these next items in this order, each in a paragraph of its own:

Immediately cite the specific position you seek. That tells him to read on. You might wish to offer *one* alternate choice, but don't list more; most places don't want a jack-of-all-trades. Add the date that you'd be available to start. Suggest a starting salary *range*. Avoid a specific sum; that could be refused. Allow room for negotiation, but don't sell yourself short; that could indicate a lack of self-esteem.

State your career goal. Organizations seek more ambitious people who have aspirations for the long haul. Toward that tack, list your special talents, skills, and professional licenses.

Tell him if you are willing to travel, or not. Do you have a valid passport? And for many technical positions in these times, are you a United States citizen by birth, or naturalized, and from where, and when?

By now you have him interested. Because it was of no relevance before the previously listed information, *now* you state your full name, address, and complete phone number. Also offer your cell phone number. Add the best times to reach you.

Enclosures: In a chart-like form, list your education, beginning with your highest degree, year, subject, and school, going back to the start of college, and include all honors and scholarships. Add extracurricular activities and athletics.

On another sheet, under a heading "Previous Employment," in a separate paragraph for each, beginning with your last (or current) job and going backward, list organization, dates from–to, position, immediate supervisor, and in a few words, describe your work (chap. 8).

On still another sheet, list your publications, and additional information such as professional societies, honors, community service, and hobbies.

Optional: You might add a professional reference person with address (after securing his permission). And if you think it might add to your stability image, you might include a *note* about your family.

End with a friendly closing (sec. 18.5), and hope for the best!

Regards.

WRITE IT AS YOU WOULD SAY IT.

19

APPENDICES

19.0

The appendix contains additional items of miscellaneous useful information and trivia germane to the overall subject, but not included in the text to avoid distraction from the narrative's objective flow.

19.1

Vide these two segments; one from a page top, the other from a page bottom. Wrong: What is *above* this page top? Right: How about *previously*? Or *earlier*?

```
                    Other Reasons Information Is Shared   ?        You can do that
e can serve you     We may share the customer data described above with     address or call
                    other companies that perform services for us or on our  process your re
                    behalf. This includes firms that provide mailing or marketing  us four to six w
                    services for us, or develop and maintain software for us.  campaigns that
Social Security     We may also share it with financial firms outside ▮▮▮▮▮,
                    such as banks or securities brokers or dealers, when we  If you ask to be
                                                                              corporate offers
ave (such as the    have agreements to jointly sponsor or offer other financial  about your poli
ccount balances,    products. We do this only if the applicable federal or state  about other prc
                    law allows the disclosure. Medical and driving record   out will not affe
```

Wrong: What is *below* this page bottom? Right: How about *next*? Or *later*?

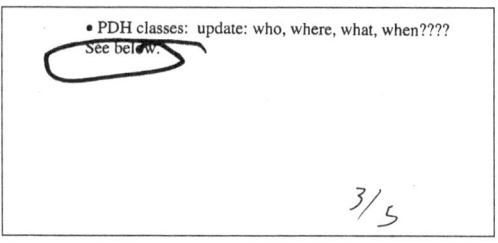

AND ALL THESE THINGS SHALL BE ADDED TO YOU.

—Matthew 6:33

APPENDIX 4.3.1 WRITE ON!—22 STYLE, FINE GROUND RULES

A while ago, in a technical writing short course for Louisiana highway people, Yvonne Lewis Day found, modified, enlarged, and then contributed this humorous list of "do as I *say*, not as I *do*" rules to aid the average writer. To hers I added a few and changed some others. Besides getting a chuckle, you might find one or two among them to catch your attention, and correct some small flaw in your own writing.

1. As you well know, it goes without saying that needless to say is needless to say.
2. Each pronoun should agree with their antecedent.
3. Verbs has to agree with their subject.
4. Remember to carefully avoid split infinitives.
5. Watch out for irregular verbs which have crope into the idiom.
6. In articles and reports we use commas to keep things apart without which we would have without doubt confusion of the most unpunctuated order.
7. But, don't use, commas, which, are not necessarily, necessary.
8. Its important to use you're apostrophe's correctly.
9. Don't use no double negatives. Not no how!
10. A writer should not shift your point of view.
11. Don't write a run-on sentence you have to punctuate it and try to do right not wrong before you write the next thought.
12. In my opinion I think that an author when he is writing something should not get accustomed to the habit of making use of too many redundant, absolutely unnecessary words that he does not actually really need in order to put his message across to the reader who is reading the article.
13. As far as incomplete constructions they are wrong.
14. About sentence fragments.
15. Just between you and I, case is important.
16. Keep it among us two, between three it's no secret.
17. About repetition, the repetition of a word is usually effective repetition over again.
18. Check carefully to if you any words left out.
19. With less dollars, you have fewer money.
20. Don't abbrev. unless nec.
21. Spele correckley.
22. A preposition is a bad thing to end a sentence with.

APPENDIX 4.3.3

A nondescript grayish fish swam in a 3" × 3" glass-faced tank at the Vancouver (BC, Canada) Aquarium. It regarded the public with the same detachment that the passersby regarded it. The plastic sign beside the tank read, *Black Sea Bass*. I asked the attendant if that were a bass from the Black Sea, or a sea bass that is black. He replied blankly that he was there just to keep people from breaking the glass, but suggested I inquire of the director. The director told me he didn't know because the fish was in place before he had arrived. Uninformed, I pondered how, by using the proper punctuation, the situation could have been demystified. *Black Sea Bass* could well have been *Sea Bass, black*; or *Bass, Black Sea*, using either punctuation properly. Some years later, an ichthyologist told me there are no sea bass in the Black Sea.

APPENDIX 4.4

Why make an easy communication difficult? Cut the bloat. The school superintendent of a Long Island, New York, district enclosed with his teachers' annual renewal letter a prepared acknowledgment form sheet to be signed, dated, and returned. It read, *This is in confirmation that I have received written notification of the continuance of my employment during the academic year of # to #*. More economically, he might have put this shortened sentence (without losing substance) on the back of a preaddressed postal card, with a blank line for signature and date: *I confirm receipt of written extension of my academic year's (# to #) employment.*

APPENDIX 4.5.1

Easily thirty years ago, Philip Broughton, a U.S. Public Health official, confected a Buzz Phrase Generator. He listed three columns of ten words each. By selecting any one word from each column, you make a three-word Buzz Phrase to drop into a report to make it appear authoritative. In Broughton's own words, "No one will have the remotest idea of what you're talking about, but . . . they're not about to admit it." Here is a report paragraph incorporating Broughton's Buzz Phrases:

> *The Computer Aided Pipe Sketching (CAPS) System integrated management concept will provide responsive organizational capacity for a total logistical contingency based upon systematized third-generation hardware. The system will ensure synchronized transitional projections combined with balanced incremental programming to which the functional policy options will be parallel. Any optional digital flexibility will be evaluated for the compatible reciprocal mobility*

to enhance this parallel monitored time-phase plan. This integrated management concept for functional organizational time-phase is total organizational flexibility. Those entities may be stored at Headquarters.

Those within a specific working area are most familiar with the working jargon and don't know or couldn't care less if outsiders know how to write a condensed style among themselves, excluding others. That same paragraph might look like this to "insiders":

The CAPS System 005 will provide 512 for a 159 based upon 288. The system will ensure 767 combined with 974 to which the 490 will be parallel. Any 641 will be evaluated for the 833 to enhance this 326 plan. Storage at HQ.

Then there are professionals concerned with advancing knowledge, who choose to express themselves in comprehensible standard English, as

Computer Aided Pipe Sketching (CAPS) will be maintained in our home office by sketchers. Input will be on a preprinted coding sheet. Use pre-edit to determine errors. Responsible sketchers will be notified of their errors; thus, they can make corrections to be included with the next day's input.

APPENDIX 4.5.2

For fun and relief, here is a Buzz Phrase Generator concocted by journalist Gregory DL Morris: This Buzz Phrase Generator can be used to pad unsatisfactorily short papers or reports, to awe your friends and associates with your vocabulary prowess, or to discover the true meaning of the three digit posted hymns of the day, or of the daily winning number. Choose any one from each column.

1. Blatant	1. Sexual	1. Coercion
2. Unilateral	2. Machiavellian	2. Leverage
3. Subconscious	3. Subliminal	3. Scenario
4. Global	4. Astronomical	4. Supply chain
5. Enhanced	5. Interpolated	5. Reengineering
6. Relational	6. Strategic	6. Partnership
7. Unconstitutional	7. Ethnocentric	7. Marketplace
8. Value-added	8. Differentiated	8. Network
9. Noncommittal	9. Individualized	9. Impact
10. Cross-functional	10. Transitional	10. Benchmark

APPENDICES   167

APPENDIX 5.1

This is the J. T. Tocci sketch of the Noyes' word picture in sec. 5.1.

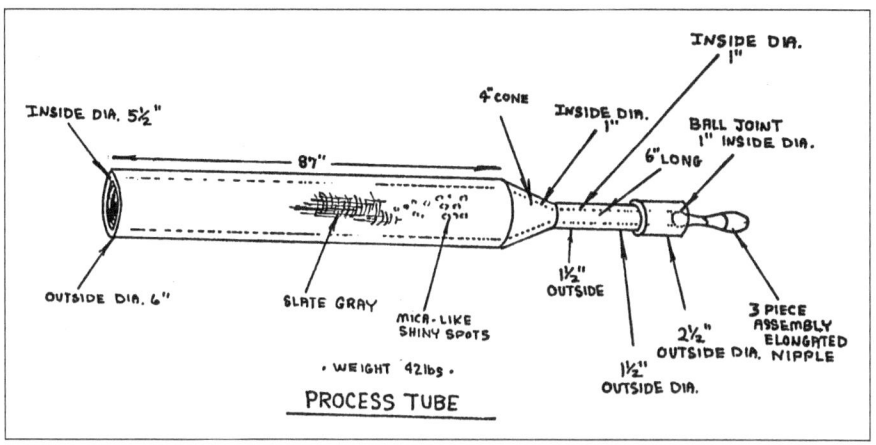

APPENDIX 5.2

Here is the sketch made from the Ted Siegal word picture in sec. 5.2

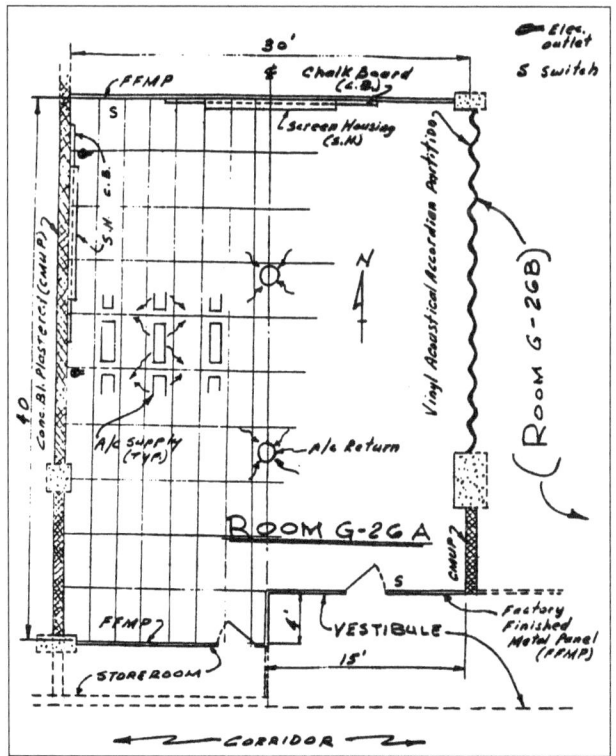

APPENDIX 7.3.15

SAFETY RULES FOR USING GRINDER (CAUTIONS ONLY)

To protect yourself, the people around you, and your equipment, you must follow these four safety rules when using the grinder:

1. Always wear safety glasses to protect your eyes from flying sparks or chips.
2. Remove rings and watches.
3. Keep sleeves above elbows to avoid catching on moving wheel.
4. Before you begin, check:

    4.1 The water pot, be sure there is sufficient water to quench the tool, otherwise the tool becomes too hot to hold.
    4.2 The tool rest, be sure it is set properly; if not, there may be a gap that can catch fingers and tools
    4.3 The wheel face, be sure it is not grooved or scored. If the wheel is damaged, fragments may break off and scatter.

Barbara Stephenson
Audiovisual Production Office
Norfolk Naval Shipyard

APPENDIX 7.3.171

(METHOD ONLY) SPECIFIC INSTRUCTIONS FOR GRINDING A SQUARE FORMING TOOL

The sequence for grinding a square forming tool is to grind (1) the end relief, (2) the back rake, finally, (3) the side relief. Cut a groove 187.5 thousandths of an inch wide and 125 thousandths of an inch deep. Begin with the tool blank and hold the tool bit to the wheel slightly above the center of the wheel, and grind the end relief. Quench the tool repeatedly in the water pot to prevent loss of temper. Now hold the top of the bit to the wheel face and grind two degrees of back rake. Next, hold the tool bit to the wheel face and grind four to six degrees of side relief on the side of the tool. As you grind the side relief, you should also grind into the body on both sides, two to four degrees of clearance from the point of the bit. Check the tool with a micrometer to be sure your dimensions are precise. Be careful not to clamp the micrometer too tightly onto the tool. Just a light feel is all that is necessary. The tool is now complete and ready to use. To review: The basic procedures, in sequence, for grinding a square forming tool are:

1. Grind the end relief,
2. Grind the back rake, and
3. Grind the side relief.

Barbara Stephenson
Audiovisual Production Office
Norfolk Naval Shipyard

APPENDIX 7.3.172

Statements not in chronological order lead to confusion. This instruction came with an old Japanese-made and marketed tape recorder. You can do the first three moves, yet it will *not* work until you do the fourth. "To load battery" belongs as the *first* instruction, to enable all else to function properly; and what is *correctly*?

How To Use

To Record:
　　Plug microphone into MICROPHONE JACK (H). Turn FUNCTION HANDLE (G) to REC. position. Adjust VOLUME CONTROL KNOB (C) to near No. 8. Recording at too high volume may produce distortion; too low volume will bring background noises to foreground.

To Rewind:
　　Turn FUNCTION HANDLE (G) to REW. position.

To Play:
　　Turn FUNCTION HANDLE (G) to PLAY position. Adjust VOLUME CONTROL (C) to desired level.

To Load Battery:
　　Remove the lid on the back, and put new batteries in correctly.

Dual Track Feature:
　　Your tape recorder is equipped to make two recordings on separate tracks of the same tape. First record on the upper half of the tape, then record on the lower half. When the tape has been used on the first track or upper half, the tape must be reversed in order to permit recording on the second track or lower half. Remember that recording over the same track or half will erase previous recording.

APPENDIX 7.4

I am moved to interject this additional gem of absurdity, which is not an instruction nor any form of motion. When a business house sends out a packet of papers, frequently it will contain one more sheet with nothing on it except this message: *This page intentionally left blank.* You'd have to be dense, really, not to detect that yourself. But once that statement is imprinted on it, that page is no longer *blank*. An efficient way to let a nervous reader know it is not a printer's goof might be to print a large X in the middle of the page.

APPENDIX 7.5 OTHER LANGUAGES

Early in the 1942 Big Band days of the ballad, Don Raye wrote his elegantly mellow number, "I'll Remember April." To have his melodic message relayed to listener's ears, composer Raye wrote his instructions in the universal musical language, to players in any orchestra. Musicians in turn, following those instructions, produced the corresponding sounds for listeners to hear, then envision the same mental impression Raye had in mind originally.

That score is the opening fragment of Raye's composition. From the top he tells his players: the clef shows the music is treble, in the key of G major, indicated by the octothorpe. Of the numbers, the top "4" calls for four beats per measure, the lower "4" demands quarter notes. The wavy line says the first beat of the first measure is a rest. The next two notes are the second and third beats after the rest. The fourth beat comes from the two eighth notes, and the eyebrow tie has the player sustain into the next measure. Try that on your piano; you might be moved to go out and get that marvelous music. (All that music technology was cheerfully given by Professor Scott Tucker of Cornell University's Department of Music.)

APPENDIX 8.2.1

An eager student once asked me how long, optimally, an article should be. I told him that was tantamount to asking me, "How heavy is a package?"

He still insisted on an answer. Taking a cue from Abraham Lincoln, who responded to a facetious question about how long a man's legs should be, with, "Long enough to reach from the ground up to his body." *Extrapolating*, I suggested, "Long enough to get your whole story told, but not so long as to put your reader to sleep."

In Jean Sibelius's dotage (1865–1957), he taught composition at a Helsinki conservatory. His advice to aspirants: "Never write an unnecessary note. Every note must live."[1] *Extrapolating*, never write unnecessary words because each word must convey its own individual piece of your idea.

APPENDIX 8.3.1 POSSIBLE SOLUTION TO SHORT-FORM DRILLS

Human tolerance to vibration has been studied under simulated ride conditions. Vibrational power and frequency were the principal characteristic input variables. Subjective judgments of ride severity are related to those input variables, but cannot be predicted numerically. Further work is required before ride severity can be predicted quantitatively. (49 words from 144)

H. B. Strock
Norton Co.,
Worcester, MA

Actually any combination of these three principal matters will do: (1) We (or "they") studied human tolerance to random vibration; (2) the predicted (designed) results, and the field (observed) results did not coincide; (3) more work is needed.

APPENDIX 8.3.6

You are writing for yourself; *the writer* is unnecessary. (Were it for someone else, it would have to be attributed). In publishing, *all* opinions should have been *studied* (word unnecessary). All *consideration* is serious. *Given by the reader* (who else)? The end colon (:) tells you, "Something follows immediately hereafter"; ergo, *the following* is unnecessary. If it is there, it is *information*, thus it is an *item*. Then the intended meaning of what I write is, *You, consider this*: But the reader is *you* and the *this* is in the listed material. Ergo, *Consider*: is the briefest complete answer. One significant word, from 24!

APPENDIX 9.2.1 REPORT FORMAT PROTOTYPE

North Atlantic Gyre Recirculation

Date: Day, Month, Year
7 pages + 3 Appendix
Unclassified

To: Dr. Cheryl Peach
   Chief Scientist
   SSV *Corwith Cramer*, Cruise 141
   Sea Education Association

From: Christopher P.M.D. Morris
   Cornell University
   SSV *Corwith Cramer*, Cruise 141
   Sea Education Association

Subject: Geostrophic Flow in the North Atlantic: In Search of Gulf Stream Recirculation.

Given: The northeastern flow of the Gulf Stream water in the North Atlantic increases from its beginnings in the Straits of Florida almost threefold to where it pushes off the continent off the Carolinas. This additional water has to come from somewhere.

Key Answer: The additional water comes from recirculation within the North Atlantic Gyre itself.

Key Terms: Geostrophic flow, gyre, Coriolis effect, Sverdrup, recirculation transport, hydrowire, rosette, hydrocast, specific volume anomaly, integrated flow, vertical pressure gradient, Ekman transport, geopotential anomaly, perpendicular transport.

Abstract: This project was designed to seek geostrophic flow recirculation from the North Atlantic Gyre system to the Gulf Stream. Depth, temperature, salinity, and density data were recorded by CTD cast at selected stations along the cruise track of the SSV *Corwith Cramer* cruise 141. We processed data to find incremental geostrophic flow and total flow for stations covering the area between casts. The study detected the presence of the Gulf Stream, subtropical convergence zone, and geostrophic recirculation.

CONTENTS

Section	Subtopic	Page
1.0	Background	1
2.0	Methodology	3
3.0	Results	5
4.0	Discussion	6
5.0	Conclusions	6
6.0	Works Cited	7
7.0	Acknowledgments	7
A	Appendix	8

<u>Abbreviations</u>: SSV—Sailing School Vessel
CTD—Conductivity, Temperature, and Density sensor
Sv—Sverdrup (amount of water transport equal to 1.0×10^6 m^3 per second)

<u>Equipment</u>: see section 2.2

<u>Conclusion</u>: see section 5.0

SIGNED
Christopher P. M. D. Morris
Cornell University
Sea Education Association

Corwith Cramer Cruise 141

GEOSTROPHIC FLOW IN THE NORTH ATLANTIC

IN SEARCH OF GULF STREAM RECIRCULATION

1.0 BACKGROUND

The image of the vast fathomless depths of the ocean, concealing mysteries for the bold adventurer to discover, has become quite cliché. But in cliché lies the seed of truth. The ocean currents have been plied by man for time out of mind. The most efficient form of propulsion was, and still is to let the water take you where it goes. Knowledge of currents and tides became among the most important secrets to glean from the sea. The ocean is an immense restless field of water on the move.

Although Alexander the Great had himself lowered in a primitive submarine vessel circa 322 BC (Allmendinger Oceanus, 1982), only in the last 100 years have the depths truly been challenged. The realm of the deep ocean belonged to the leviathans, but has become the battleground for submariners and energy companies. And as the terrestrial world becomes more and more choked with proliferating claustrophobic humans, people turn to the oceans for space, food, and power. The moats that had been formidable defenses are now highways to the front doorsteps of nations; and ultra-large crude oil carriers that draw over a hundred feet of water are just as influenced by currents as the old hide boats; more so, now that there are several city blocks of surface area to push against.

The need to understand the forces that drive, and for that matter supply water for, these currents is an environmental and an economic concern. One of the most powerful currents in the world is just off the east coast of the United States, the North Atlantic Gulf Stream (GS). "The Gulf Stream is perhaps the most widely known ocean current because of its impacts on transatlantic shipping, to influences on the European climate" (Pickart, 1994). But for all its notoriety the GS is still not fully understood.

Many of the most basic questions of cause and supply still have not been thoroughly discovered. The goal of this research is to aid in the understanding of GS transport function.

1.1 BACKGROUND RESEARCH

1.1.1 WHAT WE KNOW

The North Atlantic Ocean is home to a gigantic gyre system where water from the GS wraps across the top of the ocean basin before bending south and west back into itself. The ever-converging cycles concentrate the gyre into a "pile" of water hundreds of miles across and as much as a meter higher than the surrounding sea level. The water in this "pile" spirals downward from a combination of convergence pressure and Coriolis effect, driving the gyre from the working end.

The amount of water transported by oceanic currents is mind-bogglingly vast. The flow of the mighty Mississippi River, "the father water," is 20,000 m^3/s (Staff, Oceanus 1994). Ocean currents are measured in Sverdrups (Sv), named after the scientist who broke ground on the study in this field in the 1940s. The Sverdrup measures flow in 1,000,000 m^3/s (1.0 × 10^6 m^3/s). The powerful Mississippi barely registers on the Sverdrup scale, measuring a current of 0.02 Sv.

Consider the GS at its southern extremity, the Straits of Florida, where the North Equatorial and Guyana currents converge and flow north out of the Gulf of Mexico between Florida and Cuba. The measured flow at this position is generally accepted at around 30 Sv (Worthington 1976; Staff, Oceanus 1994). The GS then shows a rapid increase in volume as it flows north, and where it pushes off the continental coastline near Cape Hatteras, NC, the flow has been measured at anywhere between 70 and 140 Sv (Worthington 1976) (the conservative estimate falling about 80 Sv). That is a 266% increase over the Straits of Florida and more than 400,000 times the transport of the Mississippi.

1.1.2 QUESTIONS FROM PREVIOUS RESEARCH

If 30 Sv of transport is occurring through the Straits of Florida, and 80 Sv of transport is occurring in the GS off Cape Hatteras, that leaves 50 Sv unaccounted for. That missing volume should increase by 50×10^6 m/s, but this does not occur because the deficit of water is being infused into the system somewhere between Florida and North Carolina, and perhaps further north.

1.1.3 RESULTS FROM PREVIOUS RESEARCH

The most probable source for this infusion water is the great northern gyre of the North Atlantic's Sargasso Sea. But how and if water from this immense high-pressure system winds up recirculating into the GS is not known for certain. Currents at various depths indicate south and westerly transport out of the Northern Sargasso Sea (Worthington 1976).

1.2 THESIS QUESTION

The question is, where does the water come from that fills the void and keeps this vital ocean current active? Is it the Sargasso Sea? If there is an ~50 Sv flow from the Northern or Southern Sargasso Sea recirculating into the GS, then it should be detectable, as far as magnitude and direction, at depth: magnitude of volume in Sv, direction, perpendicular to station transect, along the cruise track either eastward or westward, and depth exceeding that of surface direction (wind) driven transport.

2.0 METHODOLOGY AND MATERIALS

2.1 GEOGRAPHIC REGION OF INVESTIGATION

The focus of this study was the Northern Sargasso Sea (NSS) and Southern Sargasso Sea (SSS) (approx. 41N to approx 25N). In the water column, temperature, salinity, and depth readings were taken from the surface down to bottom water, for this project 5 m to 1000 m, although recirculation transport may be occurring at lower depths. It was important to remember that data from the stations represented an average over an area between casts, rather than an integrated flow reading from the entire station. Conceivably between stations there could have been tongues of higher volume transport, or lower or even zero transport.

2.2 MATERIALS

The testing equipment we used to gather data was the conductivity, temperature, and depth meter (CTD). The Seacat CTP profiler was either lowered into the water alone off the hydrowire or mounted on the rosette as part of a hydrocast to a minimum test depth of approximately 1000 m, then retrieved. Casts were lowered when the *Cramer* was hove to, for safety, and to provide a mostly stationary monitoring platform to record flow data.

The wire was spooled out to various lengths exceeding 1000 m. That was done to compensate for wire angle. Actual maximum cast depth was not known until the profiler was winched back aboard and the data downloaded onto computer.

Once downloaded, the data were processed to usable form by the Seacat CTD software on an IBM platform. Then from this state the files were transferred to Macintosh format (Excel software) and the data parsed into delineated fields. From the parsed data fields, five columns were selected from depths of 5 m to 1000 m. Five meters was the starting point because not all casts included the data for the surface readings. The five columns were Depth (in the 5-m increment), Potential temperature, Salinity, Density, and a processed figure for Specific Volume Anomaly, based on the other data.

This field was then imparted into a preformatted spreadsheet template designed to figure geostrophic velocity then graph it versus depth increment. The template was edited to have the capacity to derive incremental volume flow and total flow for the column, for the station, the definition of the station being the area between two consecutive casts that reached a recorded depth of at least 1000 m. The spreadsheet would then graph the incremental flow versus depth, figured by subtracting the northern cast data from the southern cast data. This procedure gave results in flow, with eastward perpendicular flow as negative, and westward perpendicular flow as positive (figs. 2–9).

2.3 ANALYTICAL METHODS

To provide substantial evidence that geostrophic flow is transporting approximately 50 Sv of water out of the NSS and SSS to recirculate it into the GS along its flow to the northeast, data would be needed showing a strong (30–50 Sv) integrated flow in either a westerly or southwesterly direction.

On the surface, currents are driven by winds through Ekman transport deflecting gradually over distance because of the Coriolis effect (to the left in the northern hemisphere). But Ekman transport lasts to approximately 100 m at its deepest; below that the wind is replaced by vertical pressure gradient, expressed as force/area. Other major forces are volume, density, and gravitational acceleration. This transport through geostrophic flow, though present at the surface, theoretically extends down to the lowest reaches of the ocean, although this study stops at 100 m (Open University Course Team 1989).

The geostrophic transport relies on differences in pressure gradient, where water flows from higher to lower pressure. The pressure differential among increments of depth creates the specific volume anomaly (SVA) (measured in Pascals = kg/m·s^2). Water flows from lower average anomaly to higher, the relationship being inverse to that of pressure gradient. The distance between two monitoring points is used to find the horizontal slope pressure gradient.

But the pressure gradient increase here does not occur without Coriolis effect. Find Coriolis acceleration by:

$$2 2\Omega\, v \sin \Theta,$$

where Ω = earth's rotational constant, v = linear velocity of a body relative to the earth, Θ = latitude in degrees (Pinet 163–64).

The task was then to fill the variable gaps with data measures of depth, temperature, and salinity. Temperature and salinity are used to determine density (ρ) at specific known positions with a calculated distance between each station.

The sum of the average SVA in a column is the geopotential anomaly. We began at the basal depth of 1000 m at 0 motion and added upward toward the surface. Next we figured the dynamic height to factor in the acceleration due to gravity in the column. Then we figured velocity by subtracting the change in dynamic height from the northern station from that of the southern station, then divided by (L × 2 2Ω v sin Θ). where L = distance between casts in meters.

Next the incremental volume transport had to be figured by taking velocity and multiplying by (L × depth increment), and dividing that by 10^6 to yield a number in Sv. After that the incremental flow for the column could be summed for total flow for the station (Pond and Pickard, 1983).

The eight stations were formed by the eight CTD casts used:

Cast 012 @ 39° 36.7' N × 60° 39.2' W
Cast 018 @ 38° 15' N × 68° 44' W
Cast 024 @ 36° 17.7' N × 66° 29.3' W
Cast 028 @ 35° 6.7' N × 65° 33.2' W
Cast 037 @ 34° 00' N × 64° 20' W
Cast 040 @ 32° 57' N × 63° 21' W
Cast 049 @ 29° 41.2' N × 63° 43.6' W
Cast 054 @ 27° 9.5' N × 61° 15.1' W
Cast 060 @ 24° 25.3' N × 61° 15.1' W

(See fig. 1.)

3.0 RESULTS

The results from the data processing on the revised geostrophic flow spreadsheet were: [Editor's Note: This report continued in the same vein to conclusion.]

APPENDIX 13.6.2

Improvise when pushed to it; try a bit of humor. At an ASCE dinner I was "volunteered" to introduce the speaker, a lady who was simultaneously the mayor of a New Jersey town, and chair of the Garden State Parkway Authority. I was handed a voluminous bio sheet, and had no time to talk to her first; nor would I *ever* read from a sheet. I used the one about a woman who, on seeing a hippo submerged in a zoo pool, asked the keeper if it were male or female. His reply, "Lady, I don't think that's of any importance to anyone except another hippopotamus." To which I added, "I don't think her CV is of any importance to anyone except another P.E. who is the chair of the Garden State Parkway Authority. I am pleased to present . . . ."

APPENDIX 13.7.6

Stick to the subject. If you have a full day or longer solo presentation, resist the temptation to tell too many jokes to keep the sessions going. Occasional well-salted one-liners work, but you concentrate on making the material move, your primary objective. I taught a four-day short course on Marshall Testing to 400 asphalt contractors in Raleigh, NC. At the closing picnic, as I moved around chatting with the participants, I quickly realized they well remembered the jokes, not so the technology.

APPENDIX 15.5.5

Before you release any written effort from your desk, check it from A to Z *carefully*. Be severe with yourself *before* any editor does. Is your piece

> **Accurate:** Take care. If you're off the mark too far, you might not get another shot.
> **Brief:** !
> **Clear:** Clarity counters confusion, but don't belabor it to transparency.
> **Direct:** To kill a snake, cut off its tail close to the ears.
> **Easy:** On the eye, and on the mind. Go easy on the pen; if your readers have to hack it, they won't.
> **Factual:** Give your readers things they can use now.
> **Fanciful:** For fiction fabricators.
> **Genuine:** Deliver the goods. Nobody goes back to a phony.
> **Honest:** With the material, with the presentation, with the publication, with yourself.
> **Idiomatic:** In the language of your readers. If you "read" them, you reach them.
> **Just:** Consider all aspects of a question equally; avoid bias.

Knowledgeable: Display it, don't flaunt it.
Legitimate: Avoid cheap shots.
Marketable: If you can't sell it, you're not a professional.
Narrative: Tell a story.
Open: And tell it as it is.
Precise: In the fog, the step beyond vague is void.
Quick: In life, in wit, in quiddity.
Readable: What else?
Sincere: This keeps readers reading.
Timely: Even dossiers on dinosaurs must hook onto something current.
Understandable: The easiest thing for readers is not to read.
Valuable: Something for the readers in exchange for time spent with you from their only and irreplaceable lives.
Wise: Sage is more than turkey dressing.
Xylotomous: Shredded so nothing is hidden.
Yare: Like a good ship, smooth and easy to steer.
Zealous: Reflecting not the arrogant, but the self-esteemed energy of the writer—you! If it passes your honest eye, send it off!

APPENDIX 16.0

Groucho Marx is reputed to have quipped, "Outside of a dog, a book is man's best friend, inside a dog, it is too dark to read."[2]

APPENDIX 16.0.1

My father wisely advised me to, "Read good books and save yourself from the debasing influences of the commonplace."

APPENDIX 17.1.2

SOMETHING BETTER TO OFFER: (FROM THE 63RD ASTM ANNUAL MEETING SYMPOSIUM ON "NUCLEAR METHODS FOR MEASURING SOIL DENSITY AND MOISTURE.")

Mr. M. D. Morris (Consultant, New York, NY) (presented in written form)—Mr. Schwebel stated unequivocally that there *was* a calculated risk of radiation danger from the use of the devices in question; that there *was* high exposure to radiation when the probe was carried next to the thigh, as usual; furthermore, that the danger of mutations in future generations is a definite probability and that older men should operate these instruments to avoid such heredity transmissions.

To use the argument that we are all exposed to background and fallout radiation, and to other hazards in our daily life so why shun this added one invites danger from this calculated risk that is so easily avoided with a conventional density testing apparatus.

One panelist said that by intense testing carried out in a small area up to 50 tests per day could be made; however, that on an extended job such as a highway fill, only about 15 tests per day could be made, which is *not* more production than we get with conventional methods.

The idea that the nuclear apparatus is portable is not entirely true because its excessive weight must be transported in a jeep or pickup truck (not in a car), which in effect ties up a vehicle at all times. The initial and maintenance costs of this vehicle must be added to the nuclear testing budget. On the other hand, a cone outfit with auxiliary equipment and sand may weigh up to 80 lb, and balloon devices weigh as little as 30 lb.

An appreciable degree of specialized training must be given to the operators of nuclear devices to obtain reliable results. Conventional methods, in their simplicity and straightforwardness, are used by persons of average intelligence without such specialized instruction and cautions, and without the attendant responsibility for such intensely valuable equipment.

Nuclear devices at this time cost about $4500 initially, to which must be added maintenance, transportation, film-badge services, and certain other tangible and intangible costs. A sand-cone set sells for about $45 and balloon density outfits for about $100, while some homemade devices cost even less.

In the reported experience of the panelists there has been considerable downtime because of malfunctioning of the nuclear devices. Unless there is standby equipment, such downtime could have a serious effect on the unknown quality of work done while the equipment is broken and would be a poor explanation to any contractor forced to stop work while it was being repaired. Because of their simplicity, conventional methods do not have appreciable downtime losses, nor is standby equipment a cost nor an operational concern.

There was agreement by the panel that nuclear instruments would probably have to be recalibrated for readings in each type of soil that is tested. Considering the wide diversity of soils usually encountered on any project, this represents a considerable investment in time and money. On the other hand, conventional devices work anywhere and under any conditions.

Surface penetration of the present nuclear devices averages 4 in., according to the panel, and the feeling was that lift depths on the order of 4 to 6 in. are general. Although some earth-compaction equipment manufacturers advocate shallow lifts (and much can be said for and against each side of that argument), equipment today can compact 12-, 18-, and 24-in. lifts. A radioactive element to measure lifts of that depth from the surface would put the method out of proportion to its value from the standpoints of radiation hazards, movability, and costs.

The entire concept of nuclear density testing has concerned me since one of its earliest public airings.[3] In *Construction Methods and Equipment* (p. 211, May 1960), there is a complete comparison of nuclear with other density methods. This article, one of a series on earth compaction, is based on information from the U.S. Bureau of Public Roads, and other reputable sources, and is recommended reading. The consensus of that symposium was, nuclear methods must get more study. I advocate the development and use of any scientific moisture-density device that is more lightweight, efficient, economical, and safe; however, it is my considered opinion that the nuclear methods, as now constituted, do not represent the attainment of those goals.

APPENDIX 18.5.1 VOCABULARY ENRICHMENT

You do not have to be president to expand your word use level. Nor should you go overboard to overdo it. One new word per letter, if you use it in its correct context, is enough. Don't be ostentatious.

For my MBA course at Cornell on day one, one semester, I gave the class this A to Z list, then challenged them to use one each day in what they would write for that day's assignment. It was gratifying to see how well and creatively they used them all properly.

Try this list yourself over time, but you don't necessarily have to stay with just these; add others as you go. *Use* your dictionary.

Aleatory: Depending on a contingent event; uncertain; depending on luck, random, unorganized.
Battologize: To repeat words or phrases excessively in advertising or politics.
Captious: Characterized by a tendency to find faults, hard to please; critical, picky.
Daedal: Skillful; cleverly inventive; intricate; diversified; maze-like.
Eisegesis: A biased interpretation, especially of scripture.
Forniciform: In the form of an arched or vaulted structure.
Gnomon: The part of a sundial that casts the shadow to show the time.
Hebetude: The state of being dull, inert, and listless.
Iter: A road, thought line.
Jejune: (Food) Low in nutritive value; (Writing) lacking wisdom; immature.
Kitsch: Writing of popular, but shallow appeal; pretentious nonsense.
Lambent: Dealing gently and brilliantly with a subject.
Menticide: An organized effort to remove a person's previous opinions and replace them with radically different ones; brainwashing.

FIGURE 17.1.2. HAND ANNOTATED DRAFT

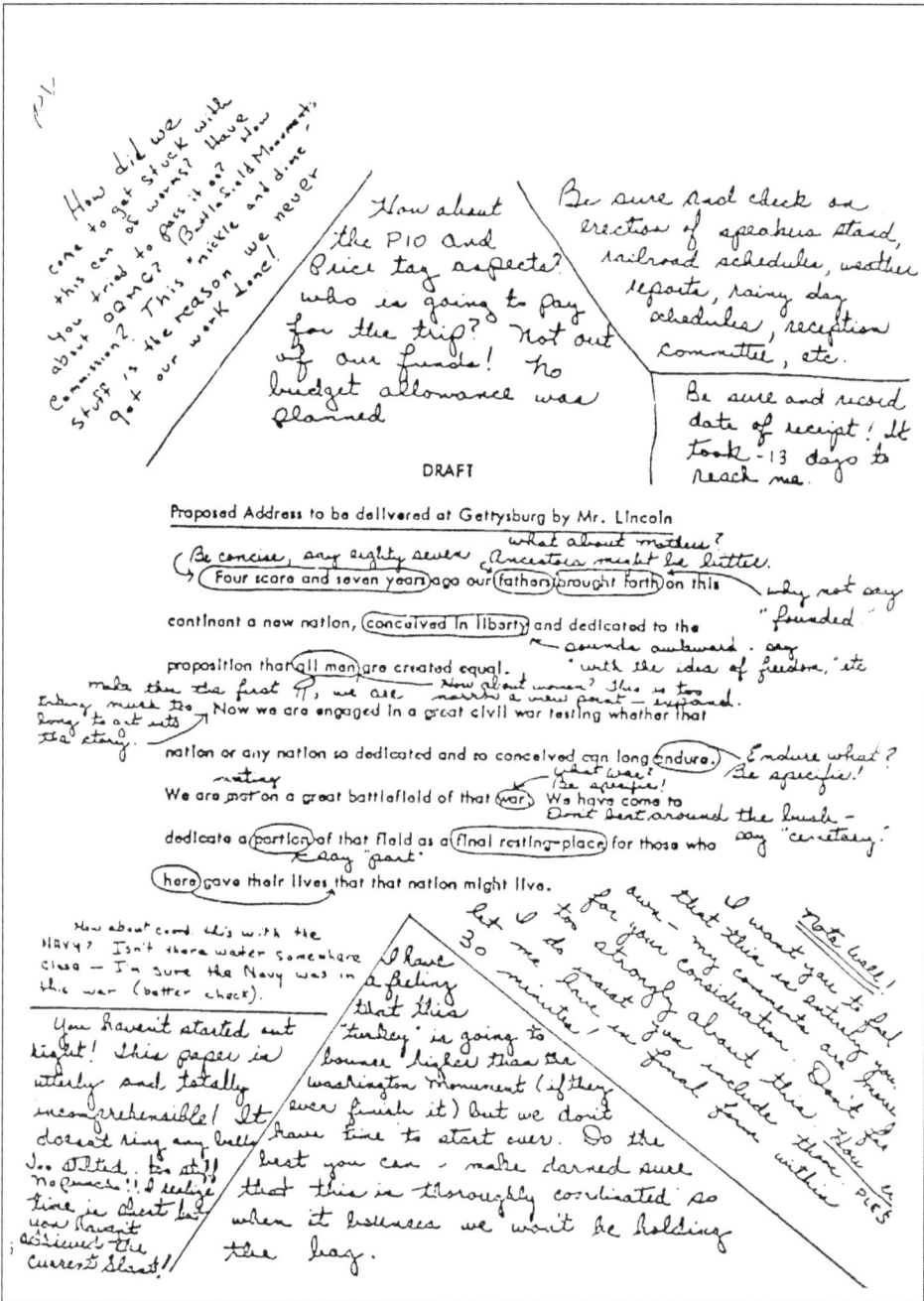

Melliflous: Sweetly flowing; smooth-sounding.
Noumenon: A thing as its true nature determines it, as contrasted with how it may appear to be.
Obdurate: Hardened against tender feelings; unyielding.
Obturate: To obstruct or stop up; close; block.
Propinquity: Closeness, nearness at the right time.
Quiddity: The essence of a thing comprehending substance and quality.
Recension: A revision of a text on the basis of a more detailed study of the sources used.
Stultify: To show in a ridiculous light; to render futile or worthless.
Triage: Selection, involving reasonably calculated risk.
Ullage: The amount by which contents, usually liquids, fall short of filling their container; unfulfilled potential.
Vagary: An erratic or unpredictable occurrence or action; a whimsical, odd, or unusual idea.
Whiffler: A man who frequently shifts his opinions, etc.; one who vacillates or is evasive in argument (sometimes waffler). (*Not* the artist who painted his mother.)
Xeric: Of, or relating to, very dry conditions, desiccated.
Yare: Quick, agile, easy to handle, as a small boat.
Zymurgy: The branch of applied chemistry concerned with fermentation (as brewing, oenology, wine-making, etc.).

This concocted, sardonic view of the old Washington bureaucracy came from the Internet, in the public domain. It conjectures what *could* happen to a good piece of work if too many hands improve something out of existence.

APPENDIX 18.5.2

More word use expansion: <u>Make</u> book you <u>make</u> out here. MAKE is a lazy writer's word. It fits more slots than a counterfeit quarter. *You* always can do better. How many different words (oddly enough, not all really synonyms) can you substitute for *make* in these 40 sentences? Do not use the same word twice, and do not change any other words. Write your answers on a separate sheet, then check *your* selections against the answers listed on the next overleaf (pg. 188).

1. I'm sorry I couldn't <u>make</u> our luncheon appointment yesterday.
2. Let us <u>make</u> an alternate lunch date please.
3. The pilot said the plane would <u>make</u> 600 miles per hour on this trip.
4. When we add this gift to the other donations, it will <u>make</u> $100,000.
5. <u>Make</u> a list of donors.
6. The law will <u>make</u> you put lights on your trailer if you travel at night.
7. Maureen buys that <u>make</u> of Scotch because it has brownie coupons on the label.
8. <u>Make</u> a solution of flour and water to use as a paste.
9. <u>Make</u> a soup of barley and beans, with beef stock.
10. If the wind holds, we should <u>make</u> port by nine o'clock tomorrow morning.
11. Christopher finds it easy to <u>make</u> friends anywhere.
12. No one knew what to <u>make</u> of the news that the fire chief quit.
13. <u>Make</u> a line for bleacher tickets.
14. The guards will <u>make</u> trouble for anyone attempting to get in free.
15. Lucius bought a large mansion that he will <u>make</u> into four apartments.
16. If you continue practicing, you will <u>make</u> an excellent gymnast.
17. Funds were allotted to <u>make</u> a bust of Ezra Cornell for Day Hall.
18. Please <u>make</u> one of your chocolate cakes for the office party.
19. One member from each school will <u>make</u> a short talk about his discipline.
20. There is just enough yarn left to <u>make</u> you a pair of winter socks.

21. His lawyer urged Wilson to <u>make</u> a will with definite bequests.
22. The contractor decided to <u>make</u> the houses on East Hill all alike.
23. Misty was happy to <u>make</u> two hundred dollars for her article.
24. Most of the tax money for defense comes from the mills that <u>make</u> steel.
25. Gregory hopes he'll <u>make</u> the finals this year.
26. The carpenter asked his boss to <u>make</u> a sketch of the bench he wanted.
27. The sales clerk promised to <u>make</u> delivery of the car by Friday.
28. The construction force will <u>make</u> a tunnel through Sagamore Mountain.
29. I <u>make</u> it my business to see that my older neighbors have transportation.
30. The factories here <u>make</u> all kinds of plastic machine parts.
31. The registrar said the course will <u>make</u> this term.
32. Everyone saw Scarlet <u>make</u> a scene at the courthouse today.
33. If you don't like it, <u>make</u> a better suggestion.
34. You can <u>make</u> a bet at the pari-mutuel window.
35. Bud won't bat this inning if you <u>make</u> out.
36. Sorry to rush, but we must <u>make</u> the last train home.
37. If Jen wins one more tournament, she'll <u>make</u> the circuit.
38. Smile, honey; you'll <u>make</u> my day.
39. We'd get by easier if you'd <u>make</u> up a good excuse.
40. <u>Make</u> sure you know what you're doing.

ANSWERS TO APPENDIX 18.5.2 MAKE EXERCISE

1. keep/honor
2. arrange/schedule/set
3. cover/go/fly
4. total/clear/net
5. compile/prepare/tally
6. insist/mandate/demand
7. brand/kind/blend
8. prepare/mix
9. cook/brew
10. reach/enter
11. meet/gain/enjoy
12. think/say
13. form/start
14. cause/impose
15. divide/renovate
16. become
17. commission
18. bake/confect
19. present/deliver
20. knit
21. write/draft/draw
22. build/construct
23. earn/receive
24. produce/fabricate
25. reach/pass/action
26. draw/draft
27. ensure/arrange/schedule
28. cut/dig/bore
29. deem/feel/find
30. produce/mold
31. run/fly
32. perform/create/set
33. offer/propose/pose
34. place/lay
35. strike/fly/ground
36. catch/board
37. complete/play
38. cheer/brighten
39. conjure/think
40. be

APPENDICES

RECAPITULATION OF ALL THE BIG IDEAS

Front Matter	Do it, don't don't it. (M. W. Jacobs, Esq.)
Chap. 1	Communication draws all (even the most barbarian) nations into civilization. (Karl Marx and Friedrich Engels)
Chap. 2	Where the needle goes, the thread will follow. (Russian proverb)
Chap. 3	There is no magic medicine of one dose. (M. D. Morris)
Chap. 4	Your finished document must reward your work. (M. D. Morris)
Chap. 5	Your word picture must create a realistic image in your reader's mind. (A. L. Morris)
Chap. 6	A good likeness will tell your reader how something looks or works, and what it may mean, or be worth. (M. D. Morris)
Chap. 7	Every body continues in its state of . . . uniform motion in a straight line. (Sir Isaac Newton)
Chap. 8	Brevity is the soul of wit. (William Shakespeare)
Chap. 9	Form ever follows function. (Louis H. Sullivan)
Chap. 10	Added material must give the reader more usable information. Every added word must work. (M. D. Morris)
Chap. 11	Govern the volume of your communication by the value of the message it carries. (M. D. Morris)
Chap. 12	To excel, a document must answer the reader's needs and wants while gratifying the professional expression of its writer. (M. D. Morris)
Chap. 13	The aim of forensic oratory is to teach, to delight, to move. (Cicero)
Chap. 14	Great spirits have always encountered violent opposition from mediocre minds. (Albert Einstein)
Chap. 15	If you can write it, you can sell it. (M. D. Morris)
Chap. 16	A good book is the best of friends, the same today and forever. (Martin F. Tupper)
Chap. 17	Criticism comes easier than craftsmanship. (Pliny the Elder)
Chap. 18	Write it as you would say it. (M. D. Morris)
Chap. 19	. . . And all these things shall be added to you. (Matthew 6:33)
Chap. 20	Knowledge is power. (Francis Bacon)
Postlude	. . . Let us exalt in the importance of ideas and information. (Edward R. Murrow)

NOTES

1. Sibelius, Jean, in radio interview with Kalevi Kilpi, 1948, http://www.sibelius.fi/english/omin_sanoin/ominsanoin_16.htm.
2. No further information is available on the source of this quote.
3. Mr. Morris' discussion of "Further Data on Gamma Ray Transmission Through Soils," by R. K. Bernhard, M. Chasek, and P. Griggs, *Proceedings, Am. Soc. Testing Mats.* 56 (1956): 1288.

20
REFERENCES AND BIBLIOGRAPHY

20.0

This book ends with a three-in-one offering. Simultaneously for this volume, here are a list of *references*, a *bibliography*, and a guide for accomplishing those two features for *your own* document.

20.1

References and bibliographies are two distinct listings serving two separate reader needs, although some reference postings also may appear in the bibliography. Both are more properly but less frequently each written as "Sources" and "Recommended Reading."

20.1.1

A list of *references* is an inventory of sources from which you *actually* have culled factual material or quotes that you used in your text. Reference citations have a different format from bibliography citations (secs. 20.2.1 and 20.2.3).

20.1.2

A *bibliography* accurately can also be called *additional reading*. In preparing that list, you enter as many items as you can find of suitable literature about the subject. You need not have read, nor even seen, any of the entries as long as you *know* they come from reliable sources and are relevant to your subject. You include a bibliography as a time-saving accommodation to your audience. His reading any entry is optional.

20.1.3 SPECIAL CASE

When the bibliography list has fewer than twenty-five entries *and* all the references are included therein, for space efficiency you may choose to combine them as I have done here. You must provide only one inventory list, the bibliography, and call out with a preceding R, those entries that are also actual references.

20.2 CONVENTION

Bibliographic entries must be uniform so any reader anywhere can read, understand, and use them. There are three major standard bibliographic citation styles in common use today.

20.2.1

American Psychological Association (APA), because of the sheer volume of publications, used mainly in psychology, education, and other social studies.

Strunk, W., Jr., & White, E. B. (2009). *The elements of style* (50th anniv. ed.). New York: Pearson Longman.

20.2.2

Modern Language Association (MLA), used principally in publications in literature, arts, and humanities.

Strunk, William, Jr., and E. B. White. *The Elements of Style*. 50th anniv. ed. New York: Pearson Longman, 2009.

20.2.3

Chicago, from the *Chicago Manual of Style* (University of Chicago), used for books, magazines, and the whole technology sphere.

Strunk, William, Jr., and E. B. White. 2009. *The elements of style*, 50th anniv. ed. New York: Pearson Longman.

I chose that last style for this work because it is used more frequently.

20.3 SOURCES AND ADDITIONAL READING (WITH REFERENCES INDICATED BY R) FOR THIS BOOK

R 1. Bartlett, John. 2002. *Bartlett's familiar quotations*, 17th ed. New York: Little Brown & Co.

R 2. Cappon, Rene J. 2000. *Associated Press guide to news writing*, 3rd ed. New York: Arco.

R 3. Dictionary of the English Language. 2005 or later. Any *reputable* publisher.

4. Dodds, Robert H. 1982. *Writing for technical and business magazines*. Malabar, FL: Krieger Publishing.
   5. Goldstein, Norm. 2007. *The Associated Press stylebook and briefing on media law*. New York: Basic Books.
   6. Harper, R. G., A. W. Wiens, and J. D. Matarazzo. 1978. *Nonverbal communication*. New York: John Wiley & Sons.
   7. Hughes, John P. 1962. *The science of language*. New York: Random House.
R 8. Kipfer, Barbara Ann. 2002. *Roget's international thesaurus*, 6th ed. New York: HarperCollins.
   9. Mitchell, John Howard. 1968. *Writing for professional and technical journals*. Ann Arbor: University of Michigan Press.
  10. Patterson, John. 1977. *Information methods*. New York: John Wiley & Sons.
  11. Pickens, Judy E. 1985. *Copy to press*. New York: John Wiley & Sons.
  12. Rosen, Harold J. 2004. *Construction specifications writing: Principles and procedures*, 5th ed. New York: John Wiley & Sons.
  13. Speake, Jennifer. 2005. *The Oxford dictionary of foreign words and phrases*. New York: Oxford University Press.
R 14. Strunk, William, Jr., and E. B. White. 2009. *The elements of style*, 50th anniv. ed. New York: Pearson Longman.
R 15. U.S. Army. 1959. *Improve your writing*. Department of the Army Pamphlet 1–10. Baltimore, MD: U.S. Army AG Publications Center.
R 16. U.S. Government Printing Office. 1987. *Word division: Supplement to United States Government Printing Office Style Manual*. Washington, DC: U.S. Government Printing Office.
  17. U.S. Government Printing Office. 2000. *Style manual*, 29th ed. Washington, DC: U.S. Government Printing Office.
  18. University of Chicago Press. 2003. *Chicago manual of style*, 15th ed. Chicago: University of Chicago Press.
R 19. Urdang, Laurence. 1985. *The New York Times everyday reader's dictionary of misunderstood, misused, and mispronounced words*. New York: Crown.

KNOWLEDGE IS POWER.

—Francis Bacon[1]

NOTE

1. 1561–1626. *Meditationes Sacrae* (1597). Reiterated by John F. Kennedy, Berkeley, CA (1962).

POSTLUDE

I close this effort with a strong idea to help you expand your universe by practicing understandable writing. This coda came from Edward R. Murrow, the man who made live, on-site radio broadcasts from wartime London. Those broadcasts were sent nightly to enable Americans to hear in real-time, enhanced word pictures of the aerial blitz. Each time, he would sign off with his own, "Good night, and good luck." More than a decade later, in 1958, in his optimistically enlightened awareness, Murrow penned these words that encapsulate my motivation for writing this book,

"LET US EXALT THE IMPORTANCE OF IDEAS AND INFORMATION."

You can, through better professional expression.

INDEX

Note: The italicized *f* following page numbers refers to figures.

A

abstracts, 82
accessible articles, 129
acknowledgments, 93, 95, 130
activation, article, 130–31
additional reading, 93, 191, 193
addressee of record, 17
advancement articles, 131–32
affirmative articles, 129
affluent articles, 129
amenities, 2
American Society of Civil Engineers (ASCE), 25, 126, 155
anecdotal articles, 129
answers, 24–25. *See also* questions
appendix, 93
applications
    formats, 93–95
    specialized formats, 112
Archimedes, 11
artful articles, 128
articles, 127–35
    acceptable, 128–29
    accessible, 129
    accurate, 129
    acknowledgments, 130
    activation, 130–31
    affirmative, 129
    affluent, 129
    anecdotal, 129
    approach and advancement, 131–32
    artful, 128
    articulate, 127
    attributes, 128–30
    ending, 132
    graphics, 132
    interview, 130
    language, 133
    marketing, 133–35
    motivation, 127
    professional, 133
    profile, 131
    readers, 127–28
    responsibilities, 132–33
    security, 132
    subject, 133
articulate articles, 129
assignment, 15
attributes, articles, 128–30
audience
    message, 5–6
    real, 6
    receivers, 17
authorship, 123–26
    chronological order, 126
    intellectual property rights, 125
    style, 153

B

Baker, K. C., 88
Barry, Bro. B. Austin, 81, 144
bibliography, 93
books, 137–47
    about, 137
    proposals, 139–42
    publishers, 137–38

books, 137–47 (*continued*)
　writing from experience, 142–47
　writing start, 138–39
brevity, 10
Buzz Phrase, 165, 166
Byron, Lord, 89

C

Cable Stayed Bridges (Scalzi), 147
captions, 42, 60
cause and effect, 83–84
chairs, 40
Chappel, J. E., 50
characteristics, writing, 42–52
　about, 43
　colloquialisms, 50–52
　complex sentences, 50
　descriptions, 56–57
　formal text, 43–44
　jargon, 50–52
　open text (publications), 44–45
　punctuation, 46–47
　redundancies, 47–48
　sentence length, 47
　spelling, 45–46
　word management, 48–49
charts, 107–9, 108*f*
Chen, Fu Hua, 144
Chi, Lu, 59
chronological order, 126
citation management programs, 25
clarity, 11–12
closings, letters, 161
colloquialisms, 50–52
columns, chart, 107
commas, 46
communication
　amenities, 2
　delivery system, 1
　growth tree, 15
　message, 1
　path direction, 11*f*
comparisons, 63–70
　about, 63
　delivery, 65

　material substance, 9
　narrative alternative, 65–70, 66*f*–69*f*
　similar entities, 64
　summary paragraph, 65
complex sentences, 50
concepts
　descriptions, 58–59
　message, 3
　motion, 73–74
conclusions, 93
Construction Contract Law (Krol), 145
Consumers Union Reports, 63
contrast, 63
control
　reason for writing, 16
　writing, 5
criticism, 5, 17
critique, 149–54
　author's style, 153
　blanket coverage, 152
　eligibility, 150
　errors, 150–51, 153
　letters to editor, 154
　praise, 152
　reasons for writing, 3, 4, 16, 150–52
　reasons not to write, 152–53
　responses or reactions, 149
　rules, 153–54
　superfluous material, 152

D

data reduction, 15, 25–27
　establish primary sort major headings, 25–26
　secondary and lesser sorts, 26–27
deadlines, 15
Debo, Harvey V., 143
decimal-numeric indexing system, 27–30, 28*f*, 44–45
　advantages, 27–28
　mistakes, 29
De Lavdibus Legum Angliae (Fortescue), 70
delivery
　communication, 1

comparisons, 65
descriptions, 61
oral presentations, 118–21
descriptions, 53–61
    about, 53
    captions, 60
    concepts, 58–59
    delivery, 61
    graphics, 59–60
    ideas, 58–59
    material substance, 8
    notions, 58–59
    receiver levels, 57–58
    style, 56–57
    writing method, 60–61
detailing, outlines, 30–31
Diamant, Leo, 143
dictionaries, 146
dimensions of message, 4f
direction, 10–11
    communication path, 11f
    origin, 10
    receipt, 10
    response, 11
    transmittal, 10
discussion. *See* critique
Donne, John, 70
drafts, writing, 41–42
dust jacket, book, 147

E

Elements of Style, The (Strunk and White), 46, 63, 192
eligibility, critique, 150
enclosures, letters, 162
ending articles, 132
end matter, 93
EndNote, 25
Engels, Friedrich, 12n1
environment, writing
    chair, 40
    implements, 40
    lighting, 40
    quiet, 39
    reference material, 40

    refreshment, 40
    timing, 40
    workplace, 39
errors, 151, 153
    critique, 150–51
    decimal-numeric indexing system, 29
Errors in Practical Measurement in Science, Engineering, and Technology (Barry), 144
ethics
    codes of, 65–67
    versus enterprise, 23
executive summary, 88–89

F

Familiar Quotations (Bartlett), 59
Fathers and Sons (Turgenev), 60
Finegan, Barton S., 50
firm outline, 37
focus analysis, 107–9
    charts, 107–9, 108f
    columns, 107
    volume versus value, 107
formal text, 43–44
formats, 91–101. *See also* specialized formats
    about, 91
    application, 93–95
    end matter, 93
    front matter, 92
    generic reports, 91–93
    letter reports, 96–98
    parallel structure, 99–100
    prototype, 94f–95f
    report body, 93
Fortescue, John, 70
Franklin, Ben, 126
front matter, 92

G

general readers formats, 113
general view of overall situation, 83
generic reports, 91–93
Geyer, Robert, 99
glossaries, 146

graphics
    articles, 132
    captions, 42, 60
    descriptions, 59–60
greetings, letters, 162
group headings, 20–21
Guidero, Elaine, 24

H

headings
    sort major, 25–26
    umbrella, 30–31, 34–35, 104, 111
higher level readers, 113
Holstrun, Lane, 96
humanistic-active style, 51*f*
hyphens, 46

I

ideas, 1, 58–59
    descriptions, 58–59
    preconceived, 12
illustrations, 93
impersonal-passive style, 51*f*
implements, writing, 40
indexing
    decimal-numeric system, 27–30, 28*f*, 44–45
    documents, 45
individual audience, 6
individual receivers, 17
influence
    reason for writing, 16
    writing, 4–5
information
    reason for writing, 16
    retrieval, 82
    writing, 4
initial unsolicited letters, 158, 161
instruction
    prototype, 75–77
    reason for writing, 16
    writing, 4
intellectual property rights, 125
interviews, 130
inventory face sheets, 24, 26*f*, 33*f*
investigation. *See* research
involvement in message, 1, 3, 10

italics, 29

J

James, Henry, 137
jargon, 50–52
job application letters, 161–62

K

keywords, 24
Klein, Richard L., 56, 79
Knapp, Brooke, 51
knowledge datum, 6, 18
Krol, John J. P., 145

L

languages
    articles, 133
    motion, 78–79
lead lines, letters, 157
letters, 155–62
    closing, 161
    to editor, 154
    enclosures, 162
    greetings, 162
    initial unsolicited, 156, 158, 161
    job application, 161–62
    lead lines, 157
    messages, 161
    model, 155–56
    openings, 159–60
    proposals, 159, 160*f*
    reports in, 96–98
    request for proposals (RFPs), 159
    responsive solicited, 158–59, 161
    rules, 158
    types, 158–59
    unsolicited proposals, 159
    writing, 155–62
lexicons, 146
Ley, Allyn G., 57
lighting, 40
Lincoln, Abraham, 122
Linnaeus, 79, 79n2
logic
    message, 3–4
    motion, 78

M

Mann, Uzi, 144
marketing, of articles, 133–35
Marx, Karl, 12n1
material substance, 5, 8–9
    comparisons, 9
    descriptions, 8
    group headings, principal matters, 20–21
    motion, 9
    pass key for projects, 22
    preliminary (provisional) outline, 19–21
    primary generic sort, 20
    questions, 19–20
    scope summary, 21–22, 22f
    superficial preresearch, 18–19
    thinking and planning (P), 18–22
McDermott, Alice, 89
McLuhan, Marshall, 1
mechanics, outline and text, 104, 105f
Men, Women and Dogs (Thurber), 47, 52n1
Merrill, Inez M., 89
messages
    apparent audience, 5–6
    communication, 1
    concepts, 3
    dimensions, 4f
    individual audience, 6
    involvement in, 1, 3, 10
    letters, 161
    logical basis, 3–4
    multiple audiences, 6–8
    primary-level audience, 8
    real audience, 6
    response to, 3
    subordinate levels, 8
    word use level, 2
mistakes. *See* errors
Momentum Press proposal guidelines, 137–40
Morris, A. L., 189
Morris, Christopher, 94, 171, 173, 195
Morris, Gregory, 125, 166
Morris, M. D., 151, 189
Morris, Misty, 49
Moss, Cheryl J., 70, 71f
motion, 73–79
    about, 73
    concept, 73–74
    instruction prototype, 75–77
    instructions, 74–75
    logic, 78
    material substance, 9
    other languages, 78–79
motivation, 4
    articles, 127
multiple audiences
    message, 6–8
    single communication, 7f
Muybridge, Eadweard, 74

N

narrative presentations
    alternative, 65–70, 66f–69f
    comparisons, 65–70
narrative style, 45
Newman, H. Michael, 143
Noyes, Gordon, 54

O

openings, letters, 159–60
open text (publications), 44–45
oral presentations, 115–22
    about, 115
    content preparation, 117
    delivery don'ts, 119–20
    delivery dos, 120–21
    delivery preparation, 118
    good, 121–22
    self preparation, 116–17
    short, 118–19
    speaker introductions, 119
organization, 13–25
    answers, 24–25
    applications, 13–14
    premise, 13
    procedure, 14f
    prototype problem, 14–15
original research, 23
outlines, 15, 29–31, 103–4
    detailing, 30–31
    enhancement, 103

outlines (*continued*)
    mechanics, 104, 105*f*
    preliminary, 19–21
    secondary sorts, 34–36
    single-entry notation, 31
    subordinated sorts, 30

P

parallel structure format, 99–100
Parlett, Pearl, 86
pass key for projects, 22
Paterson, David, 121
person piece, 130
planning. *See* thinking and planning
Pollock, Jack Harrison, 124
PowerPoint, 117
praise, 152
preconceived ideas, 12
preliminary outlines, 19–21
preparation for research, 24–25
press releases, 82
primary-level audience, 8
primary readers, 18
primary sort
    generic, 20
    major headings, 25–26
    possibilities, 34
process, short form, 85–87
professional articles, 133
professional interest areas, 112*f*
profile, articles, 131
proposals
    books, 139–42
    letters, 159, 160*f*
prototype
    comparison: narrative form, 68*f*
    instruction, 75–77
    letter report, 97–98
    parallel report structure, 99–100
    problem, the, 14–15
    report format, 92–93, 94*f*–95*f*
provisional outlines, 19–21
publishers, 138–39
punctuation, 46–47

Q

quality, short forms, 84–85
questions, 18–20. *See also* answers
quiet environment, writing, 39
quod erat demonstrandum (QED), 3

R

readers
    articles, 127–28
    higher level, 113
    involvement, 10
    primary, 18
    real, 17
rebuttal. *See* critique
recapitulation, 31–37
    determining orders of subordination, 36–37
    inventory face sheets, 33*f*
    primary sort possibilities, 34
    scope summary, 33*f*
    secondary and lesser outline sorts, 34–36
receivers, 17–18
    apparent audience, 17
    direction, 10
    individual, 17
    levels, 57–58
    preconceived ideas, 12
    primary reader, 18
    real reader, 17
    resistance, 9
    subordinate readers (secondary), 18
    thinking and planning (P), 17–18
    of writing, 5–8
recommendations, 93
recording, 5, 17
redundancies, 47–48
references, 40, 93
refinements. *See* characteristics, writing
refreshments, writing environment, 40
reports
    format prototype, 94*f*–95*f*
    formats, 93
    generic, 91–93
    in letters, 96–98
request for proposals (RFPs), 159

research, 15, 23–25
    avoid one-stop shopping, 23
    dynamic research, 23
    ethics versus enterprise, 23
    material, 37
    original, 23
    preparation, 24–25
    static, 23
responses
    critique, discussion, rebuttal, 149
    direction, 11
    to message, 3
responsibilities, article, 132–33
responsive letters, 156–57
responsive solicited letters, 158–59, 161
results, 93
resume, 161–62
Roosevelt, Franklin Delano (FDR), 125–26, 156, 161
rules
    critique, 153–54
    letters, 158
Rundlett, James R., 88

S

Scalzi, John B., 147
Schmid, Karl F., 146–47
scope summary (SS), 33*f*
    cards, 22*f*
    material substance, 21–22, 22*f*
    short forms, 87–88
secondary readers, 18
secondary sorts
    data reduction and distillation, 26–27
    outline, 34–36
security, articles, 132
sentences
    complex, 50
    length, 47
short forms, 81–90
    about, 82
    abstracts, 82
    cause and effect, 83–84
    executive summary, 88–89
    information retrieval, 82
    press releases, 82
    process, 85–87
    reduce to fewest words, 83
    retain substance and quality, 84–85
    scope summary, 87–88
    view of overall situation, 83
short oral presentations, 118–19
Sibelius, Jean, 171
Siegal, Ted, 54
signatures, 93
similar entities, comparison, 64
single communication, multiple audiences, 7*f*
single-entry notation, outlines, 31
Smetana, Bedrich, 79
Smith, Barbara L., 57
sorts
    primary, 20, 25–26
    secondary, 26–27, 34–36
    subordinated, 30
speaker introductions, 119
specialized formats, 111–13
    applications, 112
    enhancement, 111
    general readers, 113
    higher level readers, 113
    professional interest areas, 112*f*
spelling, 45–46
standard of acceptability, 9–10
    brevity, 10
    reader involvement, 10
    receiver resistance, 9
    understanding message, 9
starting writing, 40–41
static, 9
static research, 23
Stein, J. Stewart, 146
Strunk, W., 45, 63, 147
Stuyvesant High School, 82
style. *See* characteristics, writing
subject articles, 133
subordinate, 36–37
    message levels, 8
    sorts, outlines, 30
subordinate readers, 18
substance and quality, short forms, 84–85
summary paragraphs, 65

superficial preresearch, 18–19
superfluous material, 152

T

tables, 93
tape recorders, 42
targeted audiences, 37
Terkel, Studs, 23
thinking and planning, 1, 15–22
    material substance, 18–22
    reason for writing, 16–17
    receivers, 17–18
    transference, 79
    viability, 16
timing, writing environment, 40
transmittal, 10
Trollope, Anthony, 137
Tucker, Scott, 170
Turgenev, I. S., 60

U

umbrella headings, 30–31, 34–35, 104, 111
understanding messages, 9
unsolicited letters, 156–57
    initial, 158, 161
unsolicited proposals, 159

V

verbal response, 3
verbal transmittal, 10
viability, in thinking and planning, 16
volume versus value, focus analysis, 107

W

Weill, Jennifer, 121
Wen Fu (*The Art of Letters*) (Lu Chi), 59
White, E. B., 45, 63, 147
words
    management, 48–49
    use level, 2
workplace environment, 39
"World's Greatest Orations," 121–22
Writer's Digest, 133
writing
    control, 5, 16
    criticism, 5
    criticize, 17
    from experience, 142–47
    influence, 4–5, 16
    information, 4, 16
    instruction, 4, 16
    letters, 155–62
    purpose, 3–5, 150–52
    receiver of, 5–8
    record, 5, 17
writing technique, 39–42
    descriptions, 60–61
    first draft, 41
    fourth draft, 42
    optimum environment, 39–40
    second draft, 41–42
    starting, 40–41
    tape recorders, 42
    third draft, 42
    writing, 41–42

Z

Zweig, Betty, 85

www.ingramcontent.com/pod-product-compliance
Lightning Source LLC
Chambersburg PA
CBHW051643230426
43669CB00013B/2417